Robert Strauss

**Analysis of cellular resistance mechanisms to viral oncolysis**

Robert Strauss

# Analysis of cellular resistance mechanisms to viral oncolysis

Dissertation

**Südwestdeutscher Verlag für Hochschulschriften**

**Imprint**

Any brand names and product names mentioned in this book are subject to trademark, brand or patent protection and are trademarks or registered trademarks of their respective holders. The use of brand names, product names, common names, trade names, product descriptions etc. even without a particular marking in this work is in no way to be construed to mean that such names may be regarded as unrestricted in respect of trademark and brand protection legislation and could thus be used by anyone.

Publisher:
Südwestdeutscher Verlag für Hochschulschriften
is a trademark of
Dodo Books Indian Ocean Ltd., member of the OmniScriptum S.R.L Publishing group
str. A.Russo 15, of. 61, Chisinau-2068, Republic of Moldova Europe
Printed at: see last page
**ISBN: 978-3-8381-2485-8**

Zugl. / Approved by: Berlin, HU, Diss., 2010

Copyright © Robert Strauss
Copyright © 2011 Dodo Books Indian Ocean Ltd., member of the OmniScriptum S.R.L Publishing group

*"It's easy. It just needs to be done!"*

Dmitry Shayakhmetov
adenovirus researcher at the University of Washington,
who created the motto of the Lieber lab

# ABSTRACT

Vectors based on adenoviruses have been designed as targeted anti-cancer therapeutics that showed promising results in pre-clinical applications. Particularly, efforts have focused on the development of oncolytic vectors that can eliminate cancer cells and replicate in a tumor-selective fashion to amplify the input dose. In clinical trials, these oncolytic adenoviruses have generally been proved safe in patients, but have fallen short of their expected therapeutic value as monotherapies. A number of obstacles that hamper the efficacy of adenoviral vectors have been identified. These include several soluble and cellular blood components, the lack of viral receptors, and the extracellular matrix sequestered by tumor stroma. However, the apparent inability of adenoviruses to spread throughout solid tumors could not be fully explained yet.

In this thesis the susceptibility of primary ovarian cancer cells to oncolytic adenoviruses was studied in order to identify cellular mechanisms that confer resistance to virotherapy. Using gene expression profiling of cancer cells either resistant or susceptible to viral oncolysis, it was discovered that the epithelial phenotype of ovarian cancer represents a barrier to infection by commonly used oncolytic adenoviruses targeted to coxsackie- and adenovirus receptor (CAR) or CD46. Specifically, it was found that these receptors were trapped in tight junctions and not accessible for virus binding. Accessibility to viral receptors was critically linked to depolarization and the loss of tight and adherens junctions, both hallmarks of epithelial-mesenchymal transition (EMT). Importantly, tumors *in situ* as well as xenograft tumors derived from primary ovarian cancer cells mostly contained epithelial cells and cells that are in an epithelial/mesenchymal (E/M) hybrid stage when analyzed by flow cytometry and immunohistochemistry. These E/M cells are the only xenograft-derived cells that can be cultured and with passaging undergo EMT to differentiate into mesenchymal cells. Notably, only mesenchymal cells and E/M cells in the process of EMT were susceptible to viral oncolysis. On the contrary, ovarian cancer cells restricted to an epithelial phenotype conferred resistance to commonly used oncolytic adenoviruses on multiple levels. Additional resistance mechanisms, which include the activity of Rho GTPases and Rho kinase, act after successful infection by circumvention of the tight junction barrier.

In attempts to overcome the observed resistance, it was found that thus far little explored adenovirus serotypes (Ad3, Ad7, Ad11, and Ad14), which use cellular receptor(s) other than CAR and CD46, have superior oncolytic abilities on polarized epithelial tissue. These adenoviruses were able to trigger processes reminiscent of EMT in epithelial-restricted ovarian cancer cultures resulting in efficient oncolysis. This study therefore contributes to the clarification of observed discrepancies between virotherapy performances *in vitro* and *in vivo* and gives a rationale for the construction of future oncolytic adenoviruses. The observed differences in the phenotypic plasticity among cells in tumor xenografts and *in vitro* also offer new insights into the biology of cancer.

# ZUSAMMENFASSUNG

Auf Adenoviren basierende Vektoren wurden als ein gezielter Anti-Krebs-Wirkstoff entwickelt, der erfolgversprechende Resultate in prä-klinischen Studien erzielen konnte. Besonderer Wert wurde dabei auf die Herstellung von onkolytischen Vektoren gelegt, welche Krebszellen lysieren, sich tumor-spezifisch replizieren und dadurch die Ausgangsdosis erhöhen können. Solche onkolytischen Adenoviren sind zwar in klinischen Studien generell als sicher eingestuft worden, konnten jedoch als Einzelpräparat die hochgesteckten therapeutischen Erwartungen nicht erfüllen. Mehrere Hindernisse, welche die Effizienz von adenoviralen Vektoren verringern, wurden bereits identifiziert. Diese umfassen einige lösliche und zelluläre Blut-Komponenten, das Nicht-Vorhandensein von viralen Rezeptoren und die vom Tumor-Stroma abgesonderte extrazelluläre Matrix. Nichtsdestotrotz konnte dadurch die geringe Virus-Ausbreitung im Tumor-Gewebe nicht vollständig erklärt werden.

In der vorliegenden Doktorarbeit wurde die Sensitivität von primären Ovarialkarzinom-Zellen gegenüber onkolytischen Adenoviren untersucht, mit dem Ziel, zelluläre Resistenz-mechanismen zu identifizieren. Unter Verwendung der Genexpressionsprofile von Krebszellen, welche entweder resistent oder sensitiv zu viraler Onkolyse waren, konnte der epitheliale Phänotyp von Ovarialkarzinom-Zellen als Hindernis für allgemein verwendete onkolytische Adenoviren, die auf den Coxsackie- und Adenovirusrezeptor (CAR) oder CD46 ausgerichtet sind, identifiziert werden. Im Einzelnen wurde herausgefunden, dass diese zellulären Rezeptoren in der *Zonula occludens* lokalisiert und damit für Viren nicht erreichbar sind. Die Zugänglichkeit zu den Virus-Rezeptoren war zwingend an Zelldepolarisation und den Verlust der epithelialen *Zonulae occludens* und *adherens* gekoppelt, was Merkmale der Epithelial-zu-Mesenchymal-Transition (EMT) darstellt. Bedeutsam ist in diesem Zusammenhang, dass Tumore *in situ* als auch von Ovarialkarzinom-Primärmaterial abstammende Xenograft-Tumore zum größten Teil aus Epithelzellen oder epithelial/mesenchymalen (E/M) Hybrid-Zellen bestehen, wenn durch Durchfluss-Zytometrie und Immunohistochemie analysiert. Diese E/M Hybrid-Zellen sind die einzigen Zellen, welche an Zellkulturbedingungen adaptieren, wo sie durch EMT während weiterem Passagieren in Mesenchymzellen differenzieren. Bemerkenswert ist hierbei die Tatsache, dass nur Mesenchymzellen und E/M Hybrid-Zellen, die sich im EMT-Prozess befanden, sensitiv zu viraler Onkolyse waren. Im Gegensatz dazu, vermittelte der epitheliale Phänotyp von Ovarialkarzinom-Zellen Resistenz zu allgemein verwendeten onkolytischen Adenoviren auf mehreren Ebenen. Weitere Resistenzmechanismen, wie die Inaktivität von Rho-GTPasen oder der Rho-Kinase, wirkten, wenn Zellen unter Umgehung der *Zonula occludens*-Barriere erfolgreich infiziert wurden.

In Versuchen, die festgestellte Resistenz zu überwinden, wurde herausgefunden, dass bisher nur wenig erforschte Adenovirus-Serotypen (Ad3, Ad7, Ad11 und Ad14), welche einen anderen Rezeptor als CAR oder CD46 auf Zellen benutzen, besser geeignet sind, um polarisierte Epithelzellgewebe zu infizieren. Diese Adenoviren induzierten EMT-ähnliche Prozesse in Ovarialkarzinom-Kulturen mit epithelialem Phänotyp, was zu deren effizienter Onkolyse führte. Die vorliegende Arbeit trägt somit zur Aufklärung der Diskrepanz zwischen der Virustherapie-Effizienz *in vivo* und *in vitro* bei und bietet Anhaltspunkte für die Konstruktion von zukünftigen onkolytischen Adenoviren. Die unterschiedlichen phänotypischen Plastizitäten zwischen Zellen in Xenograft-Tumoren und *in vitro*-Zellkulturen geben außerdem neue Einblicke in die Biologie von Krebszellen.

# ABBREVIATIONS

| | |
|---|---|
| Ad# | Adenovirus serotype # |
| Ad#/## | Adenovirus serotype # with fiber protein of serotype ## |
| AF | Alexa Fluor |
| APC | Allophycocyanin |
| ATCC | American Type Culture Collection |
| β-gal | β-galactosidase |
| bHLH | Basic helix-loop-helix |
| bp | Base-pairs |
| CAR | Coxsackie- and adenovirus receptor |
| CB17-SCID-beige | CB-17/lcrCrl-scid-bgBR |
| CPE | *Clostridium perfringens* enterotoxin |
| COX-2 | Cyclooxygenase-2 |
| CR | Constant region |
| DNA | Deoxyribonucleic acid |
| E-cadherin | Epithelial cadherin |
| E/M | Epithelial/mesenchymal |
| EC | Extracellular cadherin |
| ECM | Extracellular matrix |
| *E.coli* | *Escherichia coli* |
| e.g. | Exempli gratia/for example |
| EGF | Epidermal growthfactor |
| EGFR | Epidermal growth factor receptor |
| EMT | Epithelial-mesenchymal transition |
| EpCAM | Epithelial cell adhesion molecule |
| EPR | Enhanced permeability and retention |
| FBS | Fetal bovine serum |
| FGF | Fibroblast growth factor |
| FGFR | Fibroblast growth factor receptor |
| Fig. | Figure |
| FIGO | Federation of Gynecology and Obstetrics |
| FITC | Fluorescein isothiocyanate |
| GFP | Green fluorescent protein |
| HGF | hepatocyte growth factor |
| HSPG | Heparan sulfate proteoglycan |
| HUGO | Human Genome Organization |
| i.e. | Id est/that is |
| i.p. | Intraperitoneally |
| i.t. | Intratumorally |
| i.v. | Intravenously |
| IFN | Interferon |
| IFP | Intratumoral fluid pressure |

| | |
|---|---|
| IGF | Insulin-like growth factor |
| IR | Inverted repeat |
| IRES | Internal ribosomal entry sequence |
| ITR | Inverted terminal repeat |
| JAM | Junctional adhesion molecules |
| kbp | Kilobase-pairs |
| MAPK | Mitogen-activated protein kinase |
| MEBM | Mammary Epithelial Basal Medium |
| MEGM | Mammary Epithelial Growth Medium |
| MET | Mesenchymal-epithelial transition |
| µg | Microgram |
| mg | Milligram |
| µl | Microliter |
| ml | Milliliter |
| µM | Micromolar |
| mM | Millimolar |
| µm | Micrometer |
| MMP | Matrix metalloprotease |
| MOI | Multiplicity of infection |
| nM | Nanomolar |
| nm | Nanometer |
| OSE | Ovarian surface epithelium |
| PBS | Phosphate-buffered saline |
| PBS-T | Phosphate-buffered saline +0.1% Tween20 |
| PE | R-Phycoerythrin |
| pfu | Plaque forming units |
| PI3K | Phosphatidylinositol 3-kinase |
| PSA | Prostate specific antigen |
| qRT-PCR | Quantitative reverse transcriptase polymerase chain reaction |
| R/E | Resistant/epithelial |
| R/E-EMT | Resistant/epithelial cells that underwent EMT |
| Rb | Retinoblastoma tumor suppressor |
| RCA | Replication competent adenovirus |
| RNA (m-, si-, mi-) | Ribonucleic acid (messenger-, small interfering-, micro-) |
| ROCK | Rho kinase |
| RSV | Rous sarcoma virus |
| RTK | Receptor tyrosine kinase |
| S/M | Susceptible/mesenchymal |
| SCID | Severe combined immunodeficiency |
| TER | Transepithelial resistance |
| TGFβ | Transforming growth factor-β |
| TRAIL | Tumor necrosis factor-related apoptosis-inducing ligand |
| VA | Virus-associated |
| ZOT | Zonula occludens toxin |

# Contents

1. **Introduction** .......... 1
   1.1. Ovarian cancer .......... 1
   1.2. Epithelial cells and their architecture .......... 2
   1.3. Epithelial-mesenchymal transition .......... 8
   1.4. Oncolytic viruses .......... 16

2. **Aim of the study** .......... 29

3. **Material and methods** .......... 31
   3.1. Material .......... 31
   3.2. Methods .......... 36
   3.3. Suppliers .......... 45

4. **Results** .......... 47
   4.1. Screening of primary ovarian cancer cultures for resistance to viral oncolysis .......... 47
   4.2. Resistant ovarian cancer cells have an epithelial phenotype .......... 49
   4.3. Adenovirus receptors are trapped within tight junctions .......... 56
   4.4. The epithelial phenotype is a barrier to adenovirus infection *in vivo* .......... 62
   4.5. The epithelial phenotype is also a barrier for adenovirus 5-based vectors .......... 66
   4.6. Pathways involved in maintenance of the epithelial phenotype .......... 68
   4.7. Adenoviruses that target receptor X trigger E-cadherin removal .......... 74

5. **Discussion** .......... 77
   5.1. Establishment and characterization of *in vitro* cultures that are resistant to viral oncolysis .......... 77
   5.2. The epithelial phenotype as a barrier for adenovirus infection and oncolysis .......... 78
   5.3. Attempts to overcome resistance to adenoviral infection and oncolysis .......... 84
   5.4. Other anatomical and physical barriers within the tumor microenvironment .......... 86
   5.5. Conclusions and future directions .......... 89

6. **References** .......... 93

7. **Supplement** .......... 117

   Acknowledgements .......... 119

This page has been intentionally left blank.

# 1. Introduction

## 1.1. Ovarian cancer

Ovarian cancer is the deadliest form of gynecological cancers in the western world (Jemal et al., 2008). About 70% of ovarian cancers are diagnosed at advanced stages due to the lack of effective screenings and the asymptomatic early phases of the disease. The 5-year survival rate for these patients does not exceed 30%. Standard therapies include surgery followed by treatment with paclitaxel and platinum-based compounds, which causes an initial response rate of 65%-80% to first-line chemotherapy (du Bois et al., 2005). However, most ovarian cancers relapse and the acquired resistance to further chemotherapy generally results in treatment failure.

Approximately 90% of primary malignant ovarian cancers are epithelial (carcinomas) and are thought to derive from ovarian surface epithelium or surface epithelial inclusion cysts (Bell, 2005; Feeley and Wells, 2001). Lately is was suggested that at least some serous ovarian carcinomas arise from the distal fallopian tube (Crum et al., 2007). In a developmental view, all of these structures originate in the same mesodermally derived embryonic coelomic epithelium, which lines the primitive peritoneal and pelvic cavities even before the ovary develops (Auersperg et al., 2008). The pluripotency of these epithelial progenitor cells might also explain the variety of structures their cancer counterparts can differentiate into. Ovarian cancers are classified entirely based on their cell morphology and subdivided into five histological subgroups: serous (about 70% of all cases), endometrioid (10%-20% of cases), mucinous (3% of cases), clear cell (10% of cases) and undifferentiated carcinomas (<1% of cases) (Seidman et al., 2004). Additionally, ovarian tumors of one cell type can be further subdivided into tumors that are benign (cystadenomas), malignant (carcinomas), or intermediate between these two (atypical proliferative tumors, tumors of low malignancy, tumors of borderline malignancy). Moreover, based on their degree of differentiation, ovarian carcinomas can be subclassified into grades 1 (well differentiated), 2 (not as well differentiated) and 3 (poorly differentiated), which correspond to <5%, 5%-50%, and >50% solid growth, respectively [FIGO (Federation of Gynecology and Obstetrics) criteria (Silverberg, 2000)]. The progression of the disease is categorized in four stages (for details see Table 1.1), ending with stage IV, which resembles the most advanced form. At this stage, tumor cells can be found in one or both ovaries and have additionally spread to parts of the body beyond the abdomen.

A series of genetic changes is thought to be the origin of ovarian cancer, but genetic signatures appear to be more specific for histological subtypes rather than for ovarian cancer in general (Aunoble et al., 2000). Additionally, the aggressiveness of tumors within certain histological types can be distinguished by genome analysis. For example, in ovarian serous carcinomas distinct altered pathways involving activation-mutated *KRAS* and *BRAF* genes or BRCA1/2 and p53 deregulation have been described for low-grade and high-grade tumors, respectively (Singer et al., 2003).

| Stage I | Cancer is found in one or both ovaries |
|---|---|
| A | Cancer is found on the inside of one ovary |
| B | Cancer is found on the inside of both ovaries |
| C | Cancer is found on the surface of one or both ovaries or in the body fluid around the ovaries |
| **Stage II** | **Cancer has spread outside the ovaries** |
| A | Cancer has spread to the uterus and/or fallopian tubes |
| B | Cancer has spread to other organs in the pelvic region (bladder, rectum, sigmoid colon) |
| C | Cancer has spread to the uterus, fallopian tubes, bladder, sigmoid colon, or rectum and may be present in the tissue or fluid of the lining of the peritoneum |
| **Stage III** | **Cancer has spread to the abdomen** |
| A | Cancer has spread to a small part of the abdomen |
| B | Cancer has spread to the peritoneum in an amount less than 2 cm |
| C | Cancer has spread to the peritoneum in an amount more than 2 cm and/or has spread to the lymph nodes |
| **Stage IV** | **Cancer has spread to parts of the body beyond the abdomen** |
| A | Cancer has spread to lungs |
| B | Cancer has spread to the liver |

**Table 1.1: Stages of epithelial ovarian cancer.** Source: National Cancer Institute, USA.
(http://www.cancer.gov/cancertopics/pdq/treatment/ovarianepithelial/Patient/page2#Keypoint10)

Multiple prognostic marker genes for ovarian cancer are reported (Heinzelmann-Schwarz et al., 2004) and characteristic genetic signatures for treatment resistance and tumor progression have been proposed (Berchuck et al., 2005; Crijns et al., 2006; Dressman et al., 2007; Hartmann et al., 2005; Spentzos et al., 2004). In summary, advanced ovarian cancers (stages III and IV) of various histological subgroups continue to be a challenging task in cancer research, because of their acquired resistance to current chemotherapy. New alternative treatment strategies, that target the metastatic stages of this disease, are therefore urgently needed.

## 1.2. Epithelial cells and their architecture

Epithelial cells constitute the epithelium, a tissue that lines the cavities and surfaces of structures throughout the body. The cells within epithelial layers form well-organized tissues by binding to each other and the surrounding extracellular matrix via specialized junctions (Fig. 1.1). Generally, epithelial cells maintain two different types of cell polarity, planar polarity and apical-basal polarity. Planar polarity is a tissue-level phenomenon that coordinates cell behavior in the two-dimensional plane of epithelial sheets (Zallen, 2007). As a consequence of the apical-basal polarity, plasma membranes of epithelial cells are divided into two specialized domains. The apical side is facing the lumen, whereas the basolateral surface connects to adjacent epithelial cells or the connective tissue. Furthermore, an asymmetrical distribution of lipids and proteins reflects different functions of these membranes. This is a result of polarized trafficking and the establishment of intercellular junctional complexes. A meshwork of proteins in tight junctions seals the paracellular space close to the apical surface, resulting in a barrier function of epithelial cell layers. Cell adhesion between neighboring cells is initiated and maintained by components of adherens junctions, which are located just underneath the tight junctions. In addition to these extracellular adhesive features, tight and adherens junctions are closely linked to the intracellular cytoskeleton. Specialized components of these complexes also play important roles in cell signaling and the regulation of gene transcription. The basal membrane of epithelial cells und the underlying connective tissue are seperated by a basal lamina, a network of extracellular matrix components including laminin, collagen IV, and fibronectin. Members of the basal lamina are secreted by epithelial cells and mainly connect to these via specific cellular integrins (Hynes, 2002).

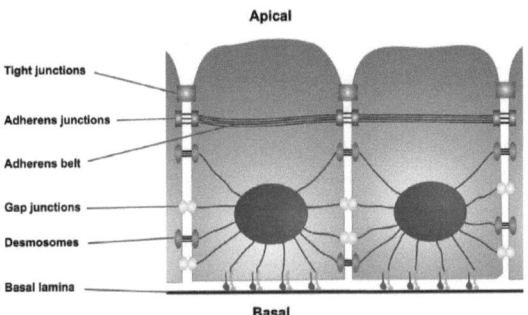

**Figure 1.1: Architecture of epithelial cells.** Adjacent epithelial cells maintain several intercellular junctions and an apical-basal polarity. Tight junctions seal the paracellular space close to the apical side. Initial cell contact is initiated by cadherins in the adherens junction complex that is situated underneath tight junctions. Adherens junction complexes encircle cells as an adherens belt, which connects to the F-actin cytoskeleton. Desmosomes are spot-like adhesions randomly arranged on lateral sides of plasma membranes. Gap junctions directly connect the cytoplasm of adjacent cells via pore complexes, allowing for free ion- and molecule-exchange. The basal lamina consists out of extracellular matrix compounds and connects to cells via cellular integrins in the basal area.

Specialized forms of epithelium can also be arranged in multilayers of epithelial cells. Within these structures, termed stratified epithelium, only one layer of cells is connected to the basal lamina, whereas the other layers adhere to each another in order to maintain structural integrity. Such epithelia are normally found in the esophagus, vagina, or oral cavity, but were also observed in several epithelial cancers including ovarian squamous cell carcinomas (Gregoire et al., 1998).

**1.2.1. Tight junctions** (zonula occludens)

Tight junctions, in combination with adherens junctions, play a key role in the formation of epithelial sheets. Strictly linked to tight junctions is a barrier function within a sheet of cells that restricts ions and small molecules to passage through the paracellular space between two adjacent epithelial cells (Madara, 1998). Additionally, tight junctions function as a 'fence' that separates the apical and basal membrane compartments in an individual cell (Turksen and Troy, 2004). Importantly, the tight junction strands on one cell are associated laterally to tight junction strands of opposing membranes on neighboring cells (Sasaki, 2003). The permselective barrier function is based on occludins and claudins, two types of transmembrane proteins that have been identified among more than 40 proteins within tight junctions (Furuse et al., 1993; Turksen and Troy, 2004). Other tight junction transmembrane proteins comprise the singlespan JAMs (junctional adhesion molecules) or CAR (coxsackie and adenovirus receptor) (Cohen et al., 2001; Furuse et al., 1993; Ikenouchi et al., 2005; Liu et al., 2000), and the lately identified tetraspan tricellulin, which is enriched in areas where three cells meet (Ikenouchi et al., 2005). None of these proteins is able to form tight junction strands on its own.

*1.2.1.1. Occludin*

The first identified integral membrane protein in tight junctions was occludin (Furuse et al., 1993), an about 65kDa tetratransmembrane protein with two extracellular loops. Two isoforms have been isolated (Muresan et al., 2000) and the localization to tight junctions is regulated by phosphorylation (Sakakibara et al., 1997). Phosphorylation and resulting trafficking to tight junctions is increased at low calcium levels (Andreeva et al., 2001). It is thought that occludin interacts in a homophilic manner with adjacent cells in order to seal the paracellular space with it's extracellular loops. However, knockout studies revealed that occludin is dispensable for tight junction formation (Saitou et al., 2000) and photoactivatable crosslinking indicated that the second extracellular loop of occludin is able to form heterophilic complexes with claudins or JAMs (Nusrat et al., 2005).

*1.2.1.2. Claudins*

Claudins are 22-27 kDa proteins that form the molecular backbone of tight junction strands. The family of claudins consists of at least 24 members in humans (Turksen and Troy, 2004). Many orthologues have been identified in multiple vertebrates and their sequence and expression analysis suggests evolutionary conserved functions (Kollmar et al., 2001). Like occludin, claudins feature four transmembrane domains and two extracellular loops, although these two protein families share no sequence homologies (Furuse et al., 1998). The first extracellular loop is partly conserved within the claudin family (Turksen and Troy, 2004), their N- and C-termini are relatively constant in lenghts and reside in the cytoplasm. Importantly, claudin-11 and -19 are reported to be capable of tight junction strand formation by themselves, based on in vitro and mouse knockout studies (Gow et al., 1999; Miyamoto et al., 2005). This highlights the core function that has been accredited to claudins in tight junctions. Supportingly, they are able to recruit occludin to tight junctions (Furuse et al., 1998). Claudins can form homophilic or heterophilic intercellular complexes in multiple combinations and it is proposed that they build pores that confer the barrier function. The ability to block the ion flow through the tight junctions can be measured as transepithelial resistance (TER), to which the individual claudins contribute differently. Claudin-1, -4, -5, -7, -8, -14 and -18 increase the TER when expressed in MDCK (Madin-Darby canine kidney) or LLC (murine Lewis lung carcinoma) cells, whereas claudin-2, -10 and -19 are able to decrease it (Tsukita et al., 2008). Most likely, these studies are highly depending on the chosen cell type, because initial TERs vary throughout different tissues. However, it is speculated that the plus and minus charges in the first loops of claudins have impact on the paracellular epithelial permeation (Furuse et al., 2001; Van Itallie et al., 2003) and are in fact responsible for anion or cation channels in tight junctions. Subsequently, the abundance of different claudins and their combinations in certain tissues is linked to the permeability of the barrier. Furthermore, altered expression of claudins has been reported for many cancers (Oliveira and Morgado-Diaz, 2007). In epithelial ovarian cancer the expression of Claudins -3, -4 and -7 is known to be elevated and positively correlates with progression of the disease (Rangel et al., 2003; Tassi et al., 2008). Claudin-3 and -4 promote migration, invasion, and survival of ovarian cancer cells and are therefore attractive candidates for targeted therepies (Agarwal et al., 2005). Additionally, these claudins have been identified as receptors for cytotoxic *Clostridium perfringens* enterotoxin (CPE). CPE binding to cells expressing claudin-3 and -4 leads to cell death and tumor growth inhibition in mice (Santin et al., 2005). Lately, a claudin-3 directed siRNA treatment also showed promising results in a mouse model for human ovarian cancer (Huang et al., 2009).

**1.2.2. Adherens junctions** (zonula adherens)

The major transmembrane proteins of adherens junctions are classical cadherins, such as epithelial cadherin (E-cadherin). Members of this protein superfamily promote homophilic intercellular adhesion in a $Ca^{2+}$-dependent manner. Their cytoplasmic domain binds cytosolic catenins that link the cadherin/catenin complex to the actin cytoskeleton. The formation of adherens junctions consequently leads to the assembly of tight junctions, but E-cadherin is dispensable for tight junction maintenance (Capaldo and Macara, 2007).

*1.2.2.1. Cadherins*

Cadherins were initially described as cell surface glycoproteins, which promote $Ca^{2+}$-dependent homophilic cell-cell adhesion (Yoshida and Takeichi, 1982). In the following years a large protein superfamily with more than 100 members in the vertebrate gene family has been identified. Despite their protein structure diversity, all cadherins share a characteristic extracellular cadherin (EC) domain (Nollet et al., 2000). Classical cadherins are the best-studied protein members of this subtype and are expressed in almost all solid tissues. They are characterized by five EC repeat domains, which are bound together by $Ca^{2+}$ ions to form rigid, rod-shaped proteins (Gumbiner, 2005; Pokutta et al., 1994). Classical cadherins can be subdivided further into types I and II. Type II cadherins are defined based on their lack of a HAV (histidine alanine valine) cell adhesion recognition sequence specific to type I cadherins. Originally, classical cadherins were named after the tissue where they are predominantly expressed, but during the past decades it became clear that the expression is not restricted to certain tissues. E-cadherin is primarily expressed in epithelial cells where it facilitates the entire epithelial junctional complex (Gumbiner et al., 1988). Additionally, epithelial cells often express P-cadherin (peritoneal cadherin) (Braga, 2000). Other classical cadherins, R-cadherin and N-cadherin (retinal and neural cadherin) are expressed in the nervous system (Matsunaga et al., 1988; Uchida et al., 1996). VE-cadherin (vascular-endothelial) lines the vasculature in the endothelium (Venkiteswaran et al., 2002) and K-cadherin (fetal kidney cadherin) is highly expressed in brain, cerebellum, and kidney. Many classical cadherins show a diverse expression pattern among different tissues and are therefore named by number (e.g. cadherin-11). Other subfamilies of cadherins comprise desmosomal cadherins, proto-cadherins, atypical cadherins and cadherin-like proteins (Gumbiner, 2005).

Classical cadherins are single-pass transmembrane proteins that form parallel (cis) homodimers with their extracellular EC domains (Fig. 1.2). Homophilic cadherin mediated cell-cell adhesion is linked to the EC1 domain (Patel et al., 2006). After initial intercellular contacts, cadherins start to cluster and then spread laterally to strengthen the contact (Adams et al., 1998; Vaezi et al., 2002). The cytoplasmic domain interacts with a number of proteins that regulate cadherin endocytosis, recycling and degradation, downstream signaling, gene transcription as well as remodeling of the underlying actin cytoskeleton (Halbleib and Nelson, 2006; Perez-Moreno and Fuchs, 2006).

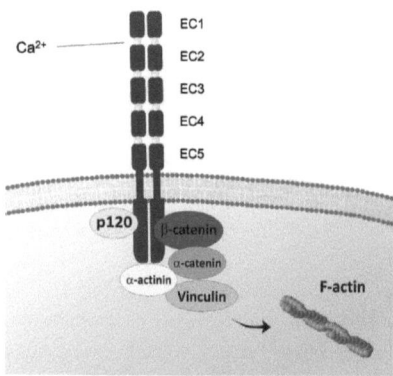

Figure 1.2: Classical cadherin/catenin complex of adherens junctions. Cadhereins form parallel homodimers. Five extracellular cadherin (EC1-5) domains, which are bound together by $Ca^{2+}$ ions, build stiff, rod-shaped proteins. The universal core cadherin/catenin complex consists of p120 catenin, bound to the junxtamembrane region, and β–catenin, bound to the distal region. The F-actin cytoskeleton connects to β–catenin (and thereby to the cadherin/catenin complex) via α–catenin, which also binds several actin-interacting proteins, as vinculin, α–actinin, or formin-1. Source: Nature Reviews Molecular Cell Biology (Gumbiner, 2005)

*1.2.2.2. E-cadherin*

The core protein of adherens junctions in epithelial cells is E-cadherin. In addition to the initiation of cell adhesion between neighboring cells, this classical cadherin functions as a key player for several cellular processes. Its cytoplasmic domain binds a number of important proteins termed catenins, which establish a connection to the actin cytoskeleton and initiate downstream signaling (Halbleib and Nelson, 2006). E-cadherin has also been directly linked to receptor tyrosine kinases (RTKs). It is co-internalized with fibroblast growth factor receptor 1 (FGFR1) after FGF stimulation (Bryant et al., 2005), suggesting a complex formation on membranes. The growth factor-mediated internalization could be counteracted by overexpression of either E-cadherin or p120 catenin in the same study. Furthermore, E-cadherin is able to negatively regulate RTK signaling by blocking their ligand-dependent activation in an adhesion dependent manner (Qian et al., 2004). This blockage is due to the extracellular domain of E-cadherin, which forms complexes with the epidermal growth factor receptor (EGFR) and is independent of β-catenin or p120 catenin binding. A recent study also addresses the effect of E-cadherin on contact inhibition and cell growth. The number of cells entering S phase was markedly reduced after homophilic ligation of recombinant E-cadherin on the cell surface. Cell growth inhibition was based on E-cadherin binding to β-catenin, but was independent of p120 catenin or binding of β-catenin to α-catenin (Perrais et al., 2007). In epithelial derived tumors the loss of intercellular adhesion is accompanied by E-cadherin down-regulation and increased cell proliferation. Restoration of E-cadherin expression in cancer cells results in decreased invasiveness, cell growth suppression and terminal differentiation (Wong and Gumbiner, 2003). It is therefore considered to function as a tumor suppressor.

*1.2.2.3. Catenins*

The catenin family comprises α-catenin, β-catenin, plakoglobin (γ-catenin), p120 catenin (δ-catenin), ARVCF (armadillo repeat gene deleted in Velo-Cardio-Facial syndrome), and p120-like catenins. Their name is the Greek word for link, initially chosen because of the connecting position between cadherins and the actin cytoskeleton (Fig. 1.2).

β-catenin is an armadillo-repeat protein, named after its homologue in Drosophila *melanogaster*, Armadillo. It is an intracellular signaling molecule involved in the Wnt pathway. When not bound to cadherins, the presence of β-catenin in the cytoplasm is low due to degradation. Upon Wnt signaling through the Frizzeled-LRP receptor the degradadion of β-catenin is inhibited and translocation to the nucleus together with T-cell factor (TCF) or leukocyte enhancing factor (LEF) occurs followed by the expression of target genes that are involved in cell proliferation (Moon et al., 2002). The affinity between E-cadherin and β-catenin is very high (Huber and Weis, 2001) and it is proposed that this interaction already occurs in the endoplasmatic reticulum, where it is required for cadherin in order to exit (Chen et al., 1999). Binding of β-catenin to E-cadherin is controlled in a phospho-regulated manner. Phosphorylation at three serine residues in the cytoplasmatic catenin binding domain of E-cadherin increases the affinity to β-catenin largely (Huber et al., 2001; Lickert et al., 2000), whereas tyrosin phosphorylation on β-catenin abolishes the binding ability to E-cadherin (Lilien et al., 2002).

α-catenin is part of the cytoskeleton and has no direct interaction with cadherins. It is thought to connect the N-terminal domain of β-catenin to actin, but the exact mechanism to date remains unclear (Yamada et al., 2005). However, it is known that α-catenin exists in a monomeric and homodimeric form. Interestingly, the α-catenin/β-catenin binding region overlaps with the α-catenin homodimerization domain on α-catenin, which excludes simultaneous binding (Pokutta and Weis, 2000). According to the newest model, α-catenin dimerizes in the cytoplasm due to concentration increase at membrane located catenin-cadherin clusters. This results in local inhibition of branching actin filament networks by competition with the Arp2/3 complex and remodels existing actin filaments to bundles instead (Drees et al., 2005).

Of special importance is the role of p120 catenin. It also features an armadillo-repeat domain but unlike β-catenin, it binds to the separate juxtamembrane domain of classical cadherins (Yap et al., 1998), which enables simultaneous binding of these two catenins to cadherin. The p120 catenin-cadherin interaction stabilizes membrane-bound cadherins and thereby increases intercellular adhesion (Davis et al., 2003; Xiao et al., 2003). Subsequently, loss of p120 catenin-induced cadherin stabilization leads to tumor progression and increased invasiveness of cancer cells (Conacci-Sorrell et al., 2002). Despite these stability functions, p120 catenin acts as an important regulator of the actin cytoskeleton through its impact on Rho family GTPases. Most cells harbor multiple isoforms of p120 catenin, which are achieved by alternative splicing of a single gene (Mo and Reynolds, 1996). N-terminal splicing results in the expression of four different ATG translation sites for p120 catenin isoforms 1, 2, 3 and 4 (Keirsebilck et al., 1998). Epithelial cells express isoforms 3 and 4, whereas mesenchymal cells mainly contain the full-length transcription variant 1 (Aho et al., 2002; Ohkubo and Ozawa, 2004). Recent studies from Panos Anastasiadis' laboratory show the diverse ways of p120 catenin signaling and its contribution to cell growth and invasiveness. They state that all isoforms of p120 catenin are capable of Rac1 activation and feature a central RhoA binding site. However, only full-length isoform 1 is able to inactivate RhoA and subsequently promote invasiveness through the regulatory N-terminal domain that is lacking in the 'epithelial' isoforms (Yanagisawa et al., 2008). Furthermore, their experiments revealed that the association with epithelial cadherin (E-cadherin) is not only critical for a sessile cellular phenotype, it also is the mediator of the tumor-suppressive function of E-cadherin by blocking Ras activation (Soto et al., 2008). In contrast, when stabilizing a mesenchymal cadherin (cadherin-11) at the plasma membrane, cell growth was induced due to p120 catenin via a Rac1-MEK1/2-ERK1/2 signaling cascade resulting in cyclin D1 activation.

To summarize, in addition to conferring a tissue barrier function for ions and solutes, epithelial intercellular junctions are closely linked to the underlying actin cytoskeleton. Additionally, several protein members of adherens or tight junctions (as β-catenin and N-p120 catenin) have important signaling functions, which can alter gene expression, cell proliferation, and the mobility of cells.

## 1.3. Epithelial-mesenchymal transition

During progression towards metastatic disease, distinct cells in epithelial cancers are thought to undergo an epithelial-mesenchymal transition (EMT), a cellular transdifferentiation program where epithelial cells lose characteristics such as tight and adherens junctions and gain properties of mesenchymal cells (Thiery and Sleeman, 2006). In contrast to epithelial cells, mesenchymal cells are defined by an irregular shape and feature unpolarized cytoskeletons and membranes. Further mesenchymal traits include increased motility, invasiveness, and elevated resistance to apoptosis. The EMT program was initially reported as a tissue culture phenomenon and its *in vivo* relevance was long subject to controversial discussions (Christiansen and Rajasekaran, 2006; Thompson et al., 2005). However, recent research on breast cancer stem cells provides strong evidence for the EMT impact in tumorigenesis (Mani et al., 2008; Morel et al., 2008). EMT enables epithelial cancer cells to leave the primary tumor, enter the bloodstream, and attach to distant organ sites in order to build metastases. This action however, involves the reversed mesenchymal-epithelial transition (MET), where cells that underwent EMT regain epithelial properties and form tumors that are histopathologically indistinguishable from the primary cancer (Brabletz et al., 2001; Thiery and Sleeman, 2006). EMT engages a series of events involving inter- and intracellular changes in affected cells. Importantly, not all of which have to occur during the transdifferantiation process. Often cells remain in stages referred to as an "incomplete" EMT, suggesting a wide spectrum of stages rather than a strict linage switch (Christiansen and Rajasekaran, 2006). EMT and MET also have a crucial role in several processes in embryogenesis. The formation of mesoderm during gastrulation or the development of several organs and structures including placenta, somites, islet cells, heart valves, and the urogenital tract all involve the interchange from epithelial to mesenchymal cell states (Christ and Ordahl, 1995; Funayama et al., 1999; Gershengorn et al., 2004; Locascio and Nieto, 2001). Moreover, in the absence of EMT, organisms other than *Porifera* and *Cnidaria*, who lack mesoderm, cannot proceed past the blastula stage during their development (Funayama et al., 1999). An important difference between the EMT in embryogenesis and the tumorigenic process lays in the genetics of the involved cells. In contrast to normal cells, cancer cells progressively lose their dependence on exogenous growth stimulation and generate many of their own growth signals (Hanahan and Weinberg, 2000). Additionally, the genome of cancer cells in large tumors is increasingly instable, which generates multiple distinct subpopulations (Loeb et al., 2008). This phenotypic diversity and the influence of the tumor microenvironment lead to a variety of ways how cancer cells might enter the EMT program.

### 1.3.1. Morphological changes during EMT

The morphological changes that occur during EMT are a consequence of diverse molecular mechanisms that contribute to the acquisition of the mesenchymal features (Fig. 1.3). A hallmark of EMT is the functional loss of E-cadherin. Subsequent breakdown of intercellular epithelial junctions plays a major role in cancer progression, where E-cadherin is therefore thought to act as a repressor of invasion (Birchmeier and Behrens, 1994). Accordingly, the reduced expression of this major regulator of the epithelial phenotype is associated with poor prognosis in several cancers (Sabbah et al., 2008). A consequent event in EMT is the shift from E-cadherin to N-cadherin (Thiery, 2002). Importantly, the homophilic intercellular junctions formed by N-cadherin are less resistant to rupture under physiological stress conditions, when compared with E-cadherin (Panorchan et al., 2006). Other cellular changes that occur during EMT include the loss of epithelial junction proteins such as claudins, occludin, desmoplakin or epithelial markers mucin-1 and EpCAM. Additionally, a change in intermediate filament proteins can be observed, where a switch from several keratins (-8, -9, -18) to vimentin occurs. The intermediate filaments contribute decisively to the mechanical rigidity of the cell. If they are composed of vimentin, the cell is flexible. Keratins, on the other hand, make the cell rigid in keeping with their function in the epithelium (Janmey et al., 1991; Wagner et al., 2007). Concomitantly with the acquisition of these mesenchymal features, the expression of several extracellular matrix proteins is induced. Fibronectin, collagen precursors, and vitronectin are all reported to be elevated in mesenchymal cells (LaGamba et al., 2005). These and other proteins up-regulated during EMT, including Src kinase, integrin-linked kinase, integrin β-5, and matrix metalloproteinases (MMPs)-11, -12, and -14, have impact on cytoskeletal remodeling and promote cell motility (Christiansen and Rajasekaran, 2006). Consequently, during a complete EMT epithelial cells lose their cobblestone-like morphology and acquire a spindle cell shape reminiscent of fibroblasts.

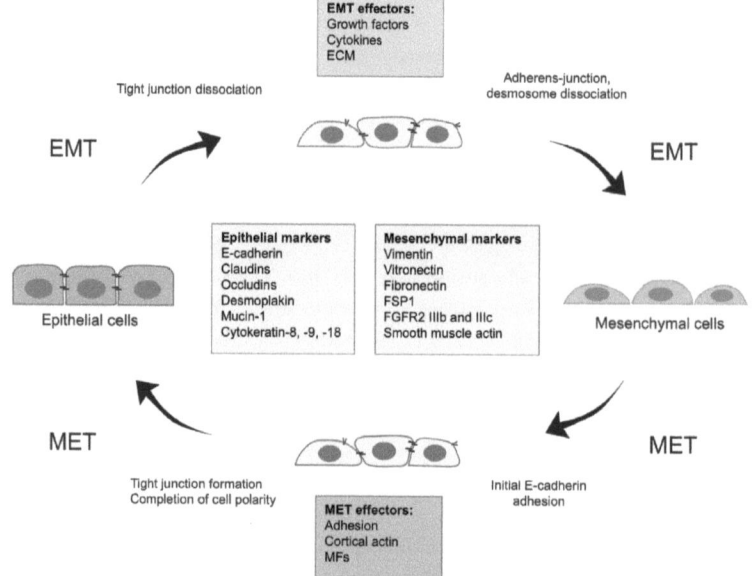

Figure 1.3: Epithelial-mesenchymal transition/Mesenchymal-epithelial transition cycle. The diagram shows the cycle of events during which epithelial cells are transformed into mesenchymal cells and vice versa. The different stages during EMT (epithelial–mesenchymal transition) and the reverse process MET (mesenchymal–epithelial transition) are regulated by effectors of EMT and MET, which influence each other. Important events during the progression of EMT and MET, including the regulation of the tight junctions and the adherens junctions, are indicated. A number of markers have been identified that are characteristic of either epithelial or mesenchymal cells and these markers are listed in BOX 1 and BOX 2. E-cadherin, epithelial cadherin; ECM, extracellular matrix; FGFR2, fibroblast-growth-factor receptor-2; FSP1, fibroblast-specific protein-1; MFs, microfilaments. Adapted from Nature Reviews Molecular Cell Biology (Thiery and Sleeman, 2006).

### 1.3.2. Induction of EMT

The EMT program is controlled on multiple levels including transcriptional repression, post-transcriptional modifications, cell signaling, and epigenetic regulation. It can be initiated by a variety of extracellular signals and specifically, the crosstalk of several responding pathways initiates a complex network, which can force cells to acquire a mesenchymal phenotype.

Moreover, sustained activation of EMT can lead to progressive epigenetic alterations, which induces inheritable effects that preserve the mesenchymal state even when EMT-inducing signals are vanished (Dumont et al., 2008).

### 1.3.2.1. Genetic and epigenetic regulation of E-cadherin

The E-cadherin encoding gene, *CDH1*, maps to a region on chromosome 16q22.1. This region is frequently associated with the loss of heterozygosity (LOH) in sporadic breast cancers (Berx et al., 1996). Furthermore, somatic mutations inactivating the *CDH1* gene are found in over 50% of diffuse-type gastric and infiltrative lobular breast cancers (Hajra and Fearon, 2002), and therefore E-cadherin has been proposed to have a fundamental role in some human cancers. However, the finding that E-cadherin mutations are rare in ductal breast cancers (Berx and Van Roy, 2001) suggests the potential involvement of epigenetic modifications that control the functions of E-cadherin. Epigenetic mechanisms such as hypermethylation of the E-cadherin promoter (Grady et al., 2000; Strathdee, 2002), Histone H3 deacetylation in the context of CpG-methylation mediated gene silencing (Koizume et al., 2002), and transcriptional silencing have all been linked to the inactivation of E-cadherin expression. The analysis of somatic cell hybrids of E-cadherin-positive and -negative breast cancer cells suggests that the loss of E-cadherin expression in some breast cancers may be linked with a dominant repression transacting pathway (Hajra et al., 1999). Overall, this supports the concept that transcriptional silencing of E-cadherin acts as a major regulatory mechanism in human cancers.

### 1.3.2.2. Transcriptional repression of E-cadherin

Several proteins have been reported to actively repress the transcription of the *CDH1* gene (Peinado et al., 2007). These transcription factors include SNAI1 (Snail), SNAI2 (Slug), ZEB1 (also known as TCF8 or δEF1), ZEB2 (also known as ZFXH1B or SIP1), E47 (also known as E2α), TCF4 (also known as E2-2) and TWIST1. SNAI1 and SNAI2 belong to the Snail protein superfamily that also includes SNAI3 and the Scratch protein family in vertebrates. Snail proteins are zinc-finger transcription factors that share a highly conserved C-terminus with 4-zinc fingers. Binding of these zinc fingers to consensus E2-boxes (CAGGTG) in the promoter region actively represses the transcription of target genes like *CDH1* (Batlle et al., 2000; Leptin, 1991; Nieto, 2002). The repressor activity of Snail proteins is critically dependent on its SNAG domain (Peinado et al., 2004). ZEB1 and ZEB2 are encoded by two independent genes and comprise the ZEB protein family. They feature two zinc finger clusters on each end, which contain 3 or 4 characteristic zinc fingers. ZEB factors interact with the DNA via simultaneous binding of the two zinc finger domains to bipartite E-boxes (CACCT and CACCTG), which are also present in the *CDH1* promoter (Comijn et al., 2001; Eger et al., 2005; Giroldi et al., 1997). Additionally, ZEB proteins can transactivate through the recruitment of either co-activators (PCAF or p300 for ZEB1) or co-repressors (CTBP for ZEB2) (Postigo et al., 2003). E47, TCF4 and TWIST1 are members of the basic helix-loop-helix (bHLH) family. The bHLH proteins also bind to the DNA using a consensus E-box (CANNTG), where they can be found as homo- or heterodimers (Ellenberger et al., 1994). Additionally to the repression of *CDH1*, bHLH members also act as direct inducers of EMT by activation of genes as N-cadherin (Perez-Moreno et al., 2001; Yang et al., 2004). Moreover, *CDH1* repressors also affect the expression of other epithelial genes. Several tight junction members including occludin, multiple members of the claudin family (-1, -3, -4, -7), and zonula occludens-3 are actively repressed in a similar manner to E-cadherin (Ikenouchi et al., 2003; Martinez-Estrada et al., 2006; Ohkubo and Ozawa, 2004; Vandewalle et al., 2005).

*1.3.2.3. EMT-regulating non-coding RNAs*

Small non-coding RNAs constitute another level of epigenetic regulation. Specifically, the recently discovered microRNAs (miRNAs) act as major post-transcriptional regulators with important functions in stem cell maintenance, differentiation and embryonic development (Filipowicz et al., 2008; Stefani and Slack, 2008). Furthermore, miRNAs can have a tumor suppressor or oncogene function (Garzon et al., 2006). Specialized RNaseIII enzymes process miRNAs into short (19-25 bp) single stranded RNAs, which are then incorporated into protein containing miRNA-induced silencing complexes. Individual miRNAs can bind to multiple mRNA targets and either induce their degradation or prevent translation. Additionally, they are reported to affect the transcription by causing methylation in the promoter region of target genes (Hawkins and Morris, 2008). Of special importance with respect to the induction of EMT are the *miR-200* family (*miR-200a, miR-200b, miR-200c, miR-141* and *miR-429*) of miRNAs and *miR-205* (Cano and Nieto, 2008; Gregory et al., 2008a; Gregory et al., 2008b). All of these miRNAs target the transcription factors ZEB1 and ZEB2, which mediate the repression of E-cadherin. Subsequently, expression of *miR-200 family* members or *miR-205* inversely correlates with a decrease in mesenchymal marker vimentin (Park et al., 2008). Furthermore, the TWIST1-induced *miR-10b* expression is associated with mesenchymal features, resulting in increased invasiveness and metastases due to elevated RhoC levels (Ma et al., 2007). In contrast, the expression of miR-335 was found to act as a suppressor of invasion and metastases (Tavazoie et al., 2008).

*1.3.2.4. Signaling pathways that regulate CDH1 repressors*

The EMT program in epithelial cells is triggered by multiple extracellular signals, which can act in a synergistic manner (Fig. 1.4). During development and carcinogenesis, EMT induced by receptor tyrosine kinases (RTKs) that are specific for certain ligands, such as fibroblast growth factor (FGF), platelet derived growth factor (PDGF), vascular endothelial growth factor (VEGF), or epidermal growth factor (EGF). Activated RTKs signal through Ras or other kinases including Src, phosphatidylinositol 3-kinase (PI3K), and mitogen-activated protein kinase (MAPK), which are all known to promote the malignant phenotype. Src activation can result in E-cadherin degradation, whereas Ras, PI3K, and MAPK signaling typically leads to induction of SNAI1 or SNAI2 (Christiansen and Rajasekaran, 2006; Thiery, 2002; Thiery and Sleeman, 2006). Furthermore, the transforming growth factor-β (TGFβ)-bone morphogenic protein pathway, or Wnt signaling are known to play important roles in EMTs occurring in embryonic development, wound healing, fibrotic diseases and cancer (Massague, 2008; Yang and Weinberg, 2008). Recently, TGFβ has also been linked to the regulation of breast cancer stem cells (Mani et al.,

2008; Morel et al., 2008) and maintenance of the pluripotent state of embryonic stem cells (James et al., 2005). The signaling mechanisms by which TGFβ induce EMT appear to be diverse and include ligand-activated receptors of SMAD transcription factors and cytoplasmic proteins that regulate cell polarity or tight junction formation (Massague, 2008; Yang and Weinberg, 2008). TGFβ receptors can phosphorylate SMAD2, SMAD3 and the cell polarity protein PAR6A, which results in loss of apical-basal polarity and concomitant disruption of epithelial tight junctions (Ozdamar et al., 2005). Additionally, TGFβ signaling through the Notch, Hedgehog, Wnt or integrin pathways can cause EMT in epithelial cells. Wnt signaling can lead to inhibition of GSK3β mediated phosphorylation of β-catenin, which prevents its subsequent degradation fasciliated by the adenomatous polyposis coli (APC) protein. This results in β-catenin accumulation and enables its translocation to the nucleus, where the protein acts as a stabilizing subunit for several EMT inducing transcription factors, including TCF4 (Caca et al., 1999; Vincan and Barker, 2008). However, the accumulation of β-catenin alone is not sufficient for EMT induction. Inactivating mutations for APC or elevated levels of β-catenin, as seen for the majority of colorectal tumors, do not necessarily correlate with mesenchymal features (Vogelstein and Kinzler, 2004). Moreover, in breast carcinomas the observed loss of E-cadherin rarely leads to nuclear β-catenin translocation. Importantly, many of the mentioned pathways can crosstalk in order to regulate SNAI1-mediated repression of E-cadherin or collaborate with β-catenin in the induction of EMT (Medici et al., 2006).

Other proteins that are known to have EMT-inducing potential through activation of SNAI1, SNAI2, TWIST1, or ZEB1 include the high mobility group protein HMGA2, endothelin 1, prostaglandin E2, stem cell factor-KIT, RAF1, calreticulin, matrix metalloprotease 3 (MMP-3), or the extracellular matrix protein laminin 5 (Peinado et al., 2007). Hormones are also reported to have positive or negative impact on the EMT-induction in several hormone-associated cancers. For example, the androgen analogue dihydrotestosterone can induce SNAI1 or SNAI2 in prostate cancer, whereas estrogen receptor-mediated signaling negatively regulates SNAI1 expression in breast cancer (Chen et al., 2006; Fujita et al., 2003). Low oxygen levels in solid tumors can also enforce EMT through multiple distinct mechanisms, including the upregulation of hypoxia-inducible factor-1α (HIF1α), hepatocyte growth factor (HGF), SNAI1 and TWIST1, NF-κB and Notch pathway activation, and induction of hypomethylation (Gort et al., 2008). Several reports show that hypoxia activated WNT-β-catenin signaling leads to SNAI1 induced EMT. The subsequent invasive tumor cell behavior is partly caused by inhibition of GSK3β-mediated β-catenin phosphorylation and destruction (Cannito et al., 2008). Additionally, the hypoxia-induced activation of SNAI1 leads to repression of E-cadherin, which induces a positive feedback loop by liberation of β-catenin from the membrane localized cadherin-catenin complex. As mentioned in the previous paragraph, this can trigger β-catenin translocation to the nucleus, where it stabilizes the expression of EMT-inducing transcription factors.

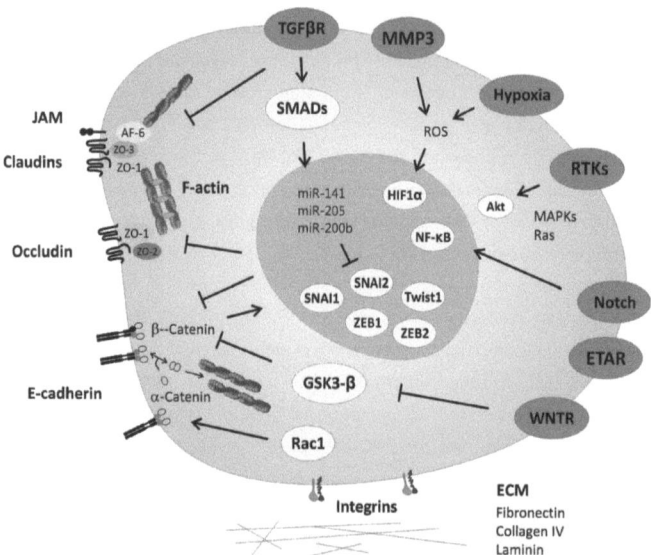

**Figure 1.4: A simplified overview of signalling networks regulating EMT.** Selected signalling pathways and some of their downstream effects and interactions are depicted. Receptor tyrosine kinases (RTKs), transforming growth factor-β (TGFβ), Notch, endothelin A receptor (ETAR), integrins, Wnt, hypoxia and matrix metalloproteinases (MMPs) can induce EMTs through multiple different signalling pathways, and the relative importance of each of these may depend on the particular cellular context. EMTs and mesenchymal–epithelial transitions (METs) are associated with dramatic changes in the cytoskeleton and extracellular matrix (ECM) composition and attachment that act together to alter cell morphology. EMT-inducing signals can lead to the disruption of tight junctions and desmosomes through protein phosphorylation (for example PAR6A phosphorylation by TGFβ signalling12) or by repressing protein levels (for example ZEB1 represses plakophilin 3). EMT also results in the dramatic reorganization of the extracellular matrix as many EMT-inducing factors upregulate the expression of ECM proteins (such as fibronectin and collagens), proteases (such as MMPs) and other remodelling enzymes (such as lysyl oxidase). Hypoxia, RAC1B activation and activation of certain kinase pathways (such as Akt) may lead to increased mitochondrial production of reactive oxygen species (ROS) that elicit pleiotropic effects, including activation of hypoxia-inducible factor 1α (HIF1α) and nuclear factor-κB (NF-κB) (orange circles), signalling and inactivation of glycogen synthase kinase-3β (GSK3β). Besides the interaction among the various signalling pathways, there is also extensive crosstalk among the EMT-inducing transcription factors (e.g. SNAI1, ZEB1) and the microRNAs (miRNAs) regulating them. E-cadherin, epithelial cadherin; H/E(Spl), hairy and enhancer of split; WNTR, Wnt receptor. Adapted from Nature Reviews Cancer (Polyak and Weinberg, 2009).

### 1.3.3. EMT and MET – Influence of the tumor microenvironment

The microenvironment within solid tumors has immense impact on the behavior of cancer cells. Specifically, interactions between epithelial and stromal cells have been linked to the induction of EMT. Importantly, in several human tumors and in animal xenograft models EMT markers are predominantly expressed at the tumor-stroma interface (Brabletz et al., 2001; Franci et al., 2006; Sheehan et al., 2008). The effect of differentiated epithelial cells on cancer cells has also been studied. In co-culture experiments with hepatocytes, prostate cancer cells underwent an MET after elevated E-cadherin expression (Yates et al., 2007). Apparently, the establishment of intercellular epithelial adherens junctions between cancer cells and hepatocytes triggered the differentiation towards the epithelial phenotype. Due to a lack of evidence, it remains questionable whether growth factors or cytokines are actively involved in the induction of MET. To date, a model where cells generally follow a default epithelial program in the absence of signals that trigger EMT seems more favorable (Frisch, 1997). A lack of tumor stroma or other extracellular signals that could activate the EMT program would therefore explain the MET that is observed after metastatic cells have reached distant organ sites (Chaffer et al., 2007).

In summary, a variety of extracellular stimuli, metabolic changes or genetic modifications can induce the EMT program in epithelial cells. Most likely the interplay of several intracellular pathways and the conditions within the tumor microenvironment determine a repression of the epithelial phenotype during disease progression, which is defined by the loss of E-cadherin and other protein members of epithelial tight and adherens junctions. The resulting morphological changes also involve the depolarization of membranes and the cytoskeleton, which leads to increased mobility and invasiveness of tumor cells.

**1.3.4. EMT in ovarian surface epithelium and ovarian cancer**

Ovarian carcinomas show a unique feature when compared to other epithelial derived cancers. Whereas the latter are characterized by the loss of epithelial properties during tumor progression, elevated expression of E-cadherin is observed for primary neoplastic ovarian epithelia (Hudson et al., 2008). The mesodermally derived normal ovarian surface epithelium (OSE) shows epithelial and mesenchymal features, characterized by the expression of both keratin and vimentin (Wong and Auersperg, 2002). Interestingly, only low levels of E-cadherin are prominent in OSE cells (Auersperg et al., 2002) and its expression is restricted to inclusion cysts and deep clefts, i.e. to areas where early malignant events are believed to occur (Sundfeldt et al., 1997). The integrity of OSE layers is primarily maintained by N-cadherin, which further highlights the epithelial/mesenchymal phenotype in this tissue (Patel et al., 2003; Peralta Soler et al., 1997). It is thought that OSE cells adapt to changes in the cellular microenvironment by transitions between epithelial and mesenchymal stages (Wong and Auersperg, 2002). Such abilities are usually restricted to immature, regenerating, or neoplastic epithelia and therefore render a unique phenotypic plasticity. It has been postulated that the reversible acquisition of a "fibroblast-like" mesenchymal cell stage during post-ovulatory repair can enhance cell motility and contractility to facilitate wound-healing, followed by matrix deposition and associated proliferative responses (Wong and Auersperg, 2002).

The initiating events in ovarian cancer remain poorly understood (Dubeau, 1999; Hudson et al., 2008), but during carcinogenesis cells acquire an increasingly complex differentiation reminiscent of the highly specialized Mullerian duct epithelium (Auersperg et al., 1994; Naora, 2007). Whereas carcinomas that arise from different tissues dedifferentiate during disease progression, ovarian tumors gain characteristics of the fallopian tube (serous carcinomas), endometrium (endometroid carcinoma), endocervix (mucinous carcinoma), and vagina (clear cell carcinoma). Generally, ovarian cancers seem to undergo an MET to EMT during progression, in contrast to the common EMT to MET that is observed in other carcinomas. However, this initial shift towards a more differentiated phenotype early in tumor progression might then be followed by a reacquisition of mesenchymal features in more advanced ovarian tumors. Interestingly, the epithelial/mesenchymal phenotype of OSE cells (Ahmed et al., 2006) is also found in the invasive front of colon (Brabletz et al., 2005) and breast cancer (Come et al., 2006), and in normal epidermal tissue during wound repair (Arnoux et al., 2005). This "hybrid" cell stage suggests that even "incomplete" EMTs can contribute to increased motility and tumor invasion. Advanced ovarian carcinomas show both absent and persistent E-cadherin expression (Darai et al., 1998; Davies et al., 1998; Peralta Soler et al., 1997). While reduced E-cadherin expression can be frequently found in late stage carcinomas and ascites derived tumor cells, the complete absence of E-cadherin is rarely observed (Cho et al., 2006; Voutilainen et al., 2006). Overall, positive E-cadherin staining is significantly decreased in samples from stage III/IV patients when compared to stage I/II tumors (Imai et al., 2004). This suggests a secondary loss of E-cadherin during progression reminiscent of EMTs in other carcinomas. Interestingly, the observed levels of N-cadherin in primary ovarian serous tumors are generally very high (about 70% of all cases) and are retained in paired metastatic lesions. In contrast, E-cadherin levels where markedly reduced in metastases when compared to primary tumors in the same study (Hudson et al., 2008).

## 1.4. Oncolytic viruses

Oncolytic viruses are defined as replicating viruses that infect and lyse cancer cells due to the cytotoxic effect of viral replication. These viruses amplify their input dose by releasing progeny virus that infects adjacent cancer cells. This process ideally continues throughout the tumor tissue until all cancer cells are eliminated. The use of viruses as anti-cancer treatments has been a therapeutic approach for almost 100 years already. The idea is based on a report from 1912 when tumor-regression after a rabies vaccination was observed in a patient with cervical cancer (DePage, 1912). Meanwhile, a broad range of viruses including adenovirus, Bunyamwera, coxsackie, dengue, feline panleukemia, Ilheus, measles, mumps, Newcastle desease, vaccinia and West Nile virus was tested for oncolytic potential in animals and humans (Asada, 1974; Gross, 1971; Huebner et al., 1956; Reichard et al., 1992; Southam and Moore, 1951). In order to restrict initial infection and viral spread to cancer cells modifications in viral genomes were necessary, but it was not until 1991 that a herpes simplex virus mutant was able to replicate in a tumor-selective fashion (Martuza et al., 1991). A few years later Frank McCormick's laboratory proposed the use of the first adenovirus mutant *dl*1520 (also known as ONYX-015) for cancer treatment, which targets cells without functional p53 for viral replication (Bischoff et al., 1996).

### 1.4.1. Adenoviruses

During the past 20 years adenoviruses were intensively studied and have evolved as one of the most commonly used gene transfer vectors in the field of gene therapy. Key features, which make adenoviral vectors an attractive vehicle for gene transfer *in vitro* and *in vivo* include the ability to easily prepare high-titer stocks of purified virus (up to $10^{13}$ pfu/ml) and the remarkable efficiency of each step in the adenovirus cell/nucleus entry process that leads to high-level gene expression. Adenoviruses are able to transduce both dividing and non-dividing cells but are mostly incapable of genome integration into host cell chromosomes. Therefore, adenovirus-based vectors have successfully been utilized to transduce a wide variety of cell types. Nevertheless, their pathogenic origin as human viruses renders these vectors immunogenic when used in *in vivo* applications.

*1.4.1.1. Adenovirus structure and life cycle*

Adenoviruses were first discovered in 1953 as agents, which spontaneously cause degeneration of primary cell cultures from human adenoid tissue (Rowe et al., 1953). Since then, 51 human serotypes of the *adenoviridae* family have been identified and divided into 4 genera (*aviadenovirus, atadenovirus, mastadenovirus, siadenovirus*) and 6 species (A-F) (de Jong et al., 2008). Adenoviruses have been shown to be responsible for a variety of illnesses including upper respiratory disease, epidemic conjunctivitis and infantile gastroenteritis (Berk, 2007). Most studies analyzing the structure of adenoviruses have been carried out with human serotypes 2 and 5. These investigations revealed that adenoviruses have an icosahedral shape (20 triangular surfaces and 12 vertices) measuring about 90nm in diameter. The virion has a protein shell (capsid) made up of 252 capsomere subunits composed of 240 hexons and 12 pentons. Each hexon is surrounded by 6 neighbouring subunits, while each penton is enclosed by 5 neighbouring subunits and has a fiber projecting from its vertex (Fig. 1.5). The adenoviral capsid harbours 4 polypeptides alongside a single copy of the double stranded DNA genome that is covalently attached at its 5' end to the terminal protein polypeptide. The genome of adenoviruses is typically around 36 kbp in length and has inverted terminal repeat (ITR) sequences of around 100-140 bp at each end. These play important roles in DNA replication as they contain viral origins of replication. A *cis*-acting packaging sequence is present within several hundred base pairs of the left hand ITR and directs interaction of the genome with encapsidating proteins. Furthermore, the genome contains 5 early transcription units (E1A, E1B, E2, E3 and E4), two delayed early units (IX and IVa2) and one major late unit that is processed to generate 5 families of late RNAs (L1-L5). It has been shown that, with the exception of E4 (Leppard, 1997), each early and late transcription unit encodes a series of polypeptides with related functions. E1A proteins are known to activate transcription and trigger entry into the S phase of the cell cycle, which renders the cell more susceptible to viral replication (Berk, 2007). Two E1B proteins interact with E1A gene products to induce cell growth (Berk, 2007). Three E2 proteins are reported to function in DNA replication (Berk, 2007), while E3 proteins mostly play a role in modulation of the anti-viral host response to adenoviruses and are therefore dispensable *for in vitro* replication (Wold et al., 1999). One of the E3 proteins facilitates efficient progeny virus release by late cytolysis of infected cells and has been named adenovirus death protein (Tollefson et al., 1996). Late proteins are either capsid components, or proteins involved in capsid assembly (Berk, 2007).

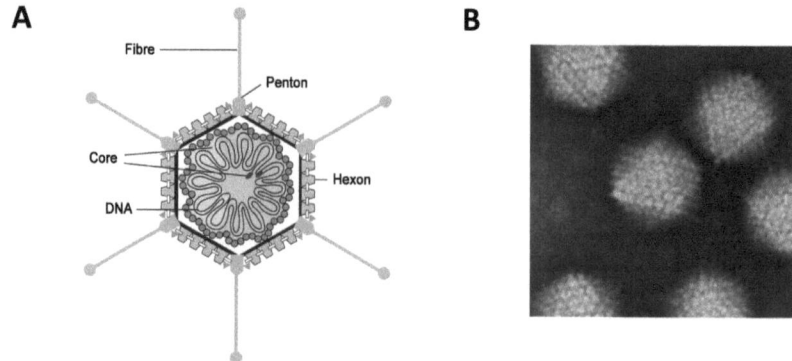

**Figure 1.5: Adenovirus structure. A)** Representative section of an adenovirus virion illustrating the current understanding of polypeptide component and DNA interactions. **B)** Electron microscopy image of adenovirus serotype 4. Adapted from Fields virology (Berk, 2007), EM picture by Daniel Stone (Stone et al., 2007).

The adenovirus life cycle (Fig. 1.6) is initiated with the binding of the adenovirus fiber knob to a high affinity cell surface receptor. Most adenovirus species, namely subgroups A and C-F, have been shown to utilize the coxsackie- and adenovirus receptor (CAR) for this purpose (Bergelson et al., 1997; Roelvink et al., 1999). The initial virus binding is followed by receptor-mediated endocytosis in clathrin-coated pits, which is mediated by interactions between an arginine-glycine-aspartic acid (RGD) motif within the viral penton base and cellular $\alpha_v b$ integrins (Wickham et al., 1994; Wickham et al., 1993). Once internalized, a drop in pH within the endosome results in a conformational change of the capsid structure, endosome disruption and release into the cytoplasm (Svensson, 1985). Hereafter, viral capsids become localized to the nucleus through a process that involves microtubules and dynein (Leopold et al., 2000). To enable this, a stepwise disassembly of adenovirus particles is necessary. This process involves fiber release, penton base dissociaton, DNA capsid scaffold protein degradation or shed, and elimination of the capsid stabilizing minor protein (Greber et al., 1993). After the capsid attaches to the nuclear pore complex, the viral DNA is injected into the nucleus (Berk, 2007) and associates with the nuclear matrix through interaction with the terminal protein (Fredman and Engler, 1993). The process of early gene transcription is initiated with the production of the viral E1A transactivator from a constitutive E1 promoter and has three major consequences. First, affected cells enter the S phase of the cell cycle to replicate the DNA. This is achieved through a number of processes including the release of E2F upon E1A binding to the retinoblastoma tumor suppressor (Rb), inhibition of the p53 tumor suppressor by E1B-55K, and direct blockage of apoptosis by the Bcl-2 homologue E1B-19K. The second consequence is the inhibition of cellular anti-viral responses, which includes the retention of major histocompatibility complex class I molecules in the endoplasmic reticulum by E3-gp19K in order to suppress cytotoxic T-lymphocyte-mediated cell death. Furthermore, single-stranded, non-coding, virus-associated

(VA) RNAs inhibit the activation of RNA-dependent protein kinase to avoid protein synthesis shut-off by the host (O'Malley et al., 1986). Recently, VA RNAs were also suggested to act as siRNAs or miRNAs that regulate viral components (Aparicio et al., 2006). The third main consequence is the synthesis of gene products needed for viral DNA replication (Berk, 2007). Following synthesis of the early gene products, DNA replication occurs within the nucleus and concomitantly the delayed early IX and IVa2 genes are transcribed. The major late promoter is activated by the IVa2 gene product and promotes production of late RNA species. Translation of late RNA species leads to production of capsid proteins within the cytoplasm, followed by their translocation to the nucleus. This ultimately results in genome packaging into assembled capsids, which are not released until the cell is lysed. Cell lysis requires disruption of intermediate filaments, such as vimentin and cytokeratin K18, which leads to collapse and rupture of the cell (Belin and Boulanger, 1987; Chen et al., 1993).

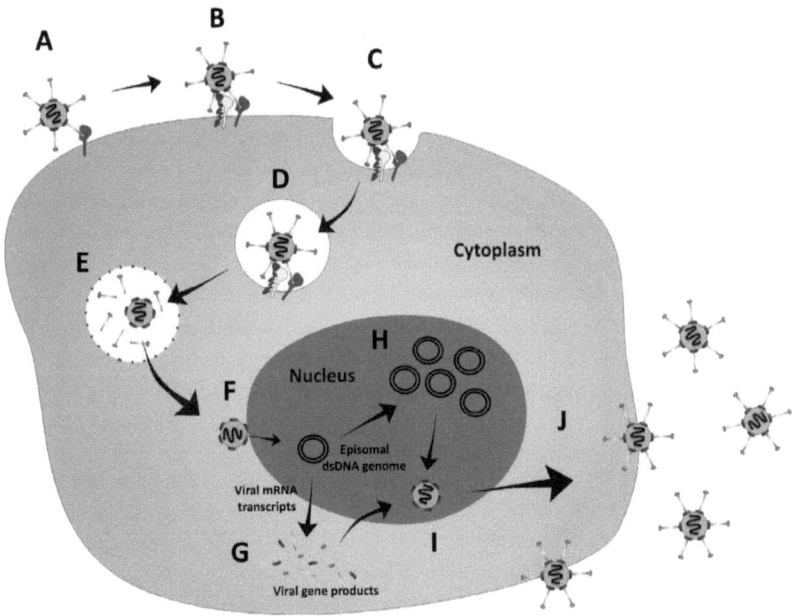

**Figure 1.6: Adenovirus life cycle.** The adenovirus knob binds to its primary receptor (CAR) (A) after which the penton base interacts with the secondary receptors ($\alpha_v b$ integrins) (B) that in turn trigger the process of endocytosis (C). Once endocytosed acidification of the endosome triggers a conformational change in the viral capsid (D) that is then released into the cytoplasm and translocates to the nucleus (F). The viral genome then enters the nucleus where it remains episomal and undergoes transcription (G) followed by replication (H). Viral gene products are then produced in the cytoplasm (G) following translation and capsid proteins localize to the nucleus where virus assembly occurs (I). Virus can then be released from the cell following lysis (J). Modified from (Stone et al., 2000).

*1.4.1.2. Development of adenoviral vectors*

The use of adenoviruses as vehicles for gene transfer was first proposed by Berkner and Sharp, who demonstrated the successful expression of dihydrofolate reductase from a 'first-generation' recombinant adenovirus (Berkner and Sharp, 1984). This 'first-generation' recombinant adenovirus lacked the adenoviral transactivator-containing E1 region and was therefore rendered replication deficient. The E1 region of this and following recombinant 'first-generation' adenoviruses was replaced by foreign DNA sequences and virus propagation must therefore occur in specific cell lines, such as HEK 293 (Graham et al., 1977), 911 (Fallaux et al., 1996) or PER.C6 (Fallaux et al., 1998), which complement the absent E1 region. This technique harbours the risk of replication competent adenovirus (RCA) occurrence, which requires careful screening of individual recombinant adenovirus batches (Fallaux et al., 1999). RCA are revertant vectors that reacquire deleted viral genes from complementing helper cells. Other 'first-generation' virus vectors may also be deleted of (Bett et al., 1995; McGrory et al., 1988) or contain an insertion (Bett et al., 1994) within the E3 coding region. The additional E3 deletion allows for insertion of up to 8 kbp foreign DNA.

'First-generation' viruses are classically generated through homologous recombination upon co-transfection of a two-plasmid system into an E1 complementing cell line. The first plasmid, termed the shuttle vector, contains the sequence of foreign DNA flanked by regions of the adenoviral genome, while the second plasmid harbours a modified version of the entire adenoviral genome. The modifications in the adenovirus genome prevent packaging by either deletion of the packaging signal or insertion of non-viral sequences that render the genome too large to be packaged. Following homologous recombination a packageable, E1/E3 deficient recombinant genome is generated. Alternative strategies, in which the adenovirus genome is manipulated in *Escherichia coli* (*E.coli*) (Chartier et al., 1996; Crouzet et al., 1997) or yeast (Ketner et al., 1994), have also been developed in order to create functional virus in eukaryotic cells.

Further crippling of the adenovirus genome led to the development of 'second-generation' recombinant adenoviruses. These vectors typically carry deletions in E2 or E4 in addition to deleted E1 and E3 regions and allow increased cloning capacity. 'Second-generation' vectors offer reduced viral gene expression, additional blocks to viral replication in transduced tissue, and reduced incidence of RCA formation, due to an increased number of individual recombinations needed for a replication competent genome. Overall, this renders these vectors less immunogenic than 'first generation' adenoviruses (Shen, 2007). Adenovirus based vectors that are devoid of all viral coding sequences have also been developed (Fisher et al., 1996; Kochanek et al., 1996; Parks et al., 1996). These vectors have been termed 'gutless', helper-dependent or high capacity adenoviral vectors and contain only the viral ITR and packaging sequences, along with the transgene coding sequences and stuffer DNA. Propagation can only occur in the presence of a helper virus that provides all proteins necessary for successful packaging of an adenovirus-based vector *in trans*.

## 1.4.2. Tumor-targeted adenoviruses

Tumor targeting of adenoviruses can be achieved using three distinct strategies: *i)* transductional targeting, *ii)* transcriptional targeting, or *iii)* the generation of conditionally replicating viruses. Transductional targeting directs the viral infection tropism to proteins that are uniquely expressed or up-regulated on the cancer cell surface. Additionally, the vectors' infectious potential for non-tumor cells must be minimized. Transcriptional targeting restricts the expression of therapeutic transgenes to tumor cells, typically using promoters that are tissue-specific or uniformly activated in cancer cells. Conditionally replicating adenoviruses infect cancer cells, specifically replicate in them and lyse them while releasing progeny virus.

### *1.4.2.1. Transductional tumor targeting*

A major task in cancer virotherapy is to achieve targeted infection of metastatic tumors after intravenous application of adenoviral vectors. In the past, vectors based on seroptype 5 (Ad5) have been used for *in vivo* gene transfer. However, Ad5 vectors predominantly transduce hepatocytes after intravenous injection, and because tumor cells often do not express the Ad5 receptor, CAR, these vectors are suboptimal for tumor targeting (Bergelson et al., 1997a; Li et al., 1999; Miller et al., 1998; Okegawa et al., 2000). Several strategies have been exploited towards achieving tumor-targeted infection with Ad5 vectors, including the genetic modification of Ad5 by incorporation of peptide motifs into specific sites of the viral capsid, the complete substitution of Ad5 fiber with heterologous targeting moieties as well as chemical Ad5 capsid modification (Campos and Barry, 2007). These approaches succeeded in changing the tropism of Ad5 to receptors that are predominantly expressed on tumor cells, such as EGFR, FGFR, Her2/neu, and specific integrins. Another strategy for redirecting Ad5 vectors to tumor cells exploits the natural diversity of the adenoviridae family. Previous work has shown that the Ad5 fiber or the C-terminal fiber knob domain can be swapped with fibers or fiber knobs of other adenovirus serotypes thereby creating "pseudotyped" hybrid vectors (Krasnykh et al., 1996; Stevenson et al., 1997) (Fig. 1.7). While most of the adenovirus serotypes utilize CAR as a primary attachment receptor, this is not the case for species B adenovirus serotypes. Species B adenoviruses form two genetic clusters, B1 (Ad3, Ad7, Ad16, Ad21, and Ad50) and B2 (Ad11p, Ad14, Ad34, and Ad35). Recently, a new grouping of species B adenoviruses based on their receptor usage was suggested (Gustafsson et al., 2006; Marttila et al., 2005; Segerman et al., 2003a; Segerman et al., 2003b; Tuve et al., 2006). Group 1: (Ad16, 21, 35, 50) nearly exclusively utilizes CD46 as a receptor; Group 2: (Ad3, Ad7, 14) shares the same non-identified receptor/s, which was/were tentatively named receptor X; Group 3: (Ad11p) preferentially interacts with CD46, but also utilizes receptor X if CD46 is blocked. Although in humans CD46 is expressed on all nucleated cells at a low level, RNA and protein studies with biopsy material have shown that CD46 expression is greatly upregulated in malignant tumor cells, including breast, colon, liver and endometrial cancers (Fishelson et al., 2003; Kinugasa et al., 1999; Murray et al., 2000). This makes Ad5 vectors containing fibers or fiber knobs derived from species B group 1 adenoviruses interesting for tumor gene therapy. Studies in non-human primates and CD46-transgenic mice that express CD46 in a pattern and at a level similar to humans demonstrated that hepatocyte transduction was significantly less pronounced with Ad5/35 vectors compared to Ad5 vectors (Ni et al., 2005; Ni et al., 2006). In models with pre-established liver metastases, intravenously injected Ad5/35 vectors achieved tumor-localized transgene expression, however the transduction efficiency of tumor cells was generally less than 5% (Ni et al., 2006). While the receptor of species B group 2 adenoviruses is still elusive, numerous studies with Ad5 vectors containing Ad3 fibers or fiber knobs have shown superior tumor cell transduction *in vitro* and in vivo (Kanerva et al., 2002; Kangasniemi et al., 2006).

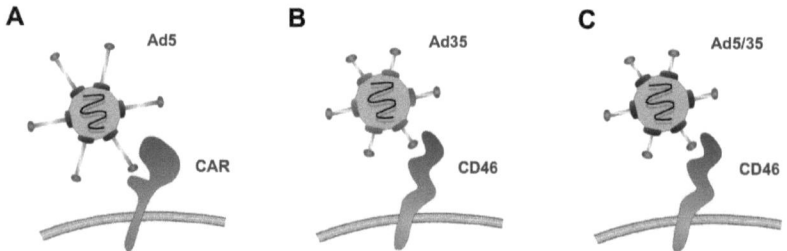

**Figure 1.7: Pseudotyping of adenoviruses. A)** The fiber of adenovirus serotype 5 (Ad5) interacts with CAR for initial cell attachment. **B)** Adenovirus serotype 35 (Ad35) attaches to cells via fiber-binding to CD46. **C)** The incorporation of the serotype 35 fiber into adenovirus 5 (Ad5/35) re-targets these vectors to CD46.

*1.4.2.2. Transcriptional tumor targeting*

Transcriptional targeting has been employed to achieve tumor-specific expression of therapeutic genes with 'first generation' adenoviruses. Several promoters that are known to be specifically active in certain cancers were used for this purpose. These targeting-approaches include promoters of cyclooxygenase-2 (COX-2) for gastrointestinal cancer (Yamamoto et al., 2001), carcinoembryonic antigen (CEA) for colorectal liver metastases (Brand et al., 1998), and human glandular kallikrein 2 (hK2) for prostate cancer (Xie et al., 2001). Other promoters that have been used in adenoviral vectors target a more wide range of cancers. Examples are promoters of the hypoxia inducible factor (HIF) for expression in hypoxic cells (Post and Van Meir, 2003), of E2F (Tsukuda et al., 2002), and of human telomerase reverse transcriptase (hTERT) (Fujiwara et al., 2007).

*1.4.2.3. Conditionally replicating adenoviruses*

Several strategies for tumor-selective adenoviral replication have been evaluated, including the deletion of adenovirus E1A and/or E1B functions or the regulation of E1A and/or E1B expression with tumor-specific promoters. One of the first oncolytic adenoviruses was *dl*1520, an Ad2/5 chimera, which lacks functional E1B-55K (Barker and Berk, 1987). As described in

1.4.1.1., the expression of E1A forces adenovirus-infected cells into S phase via Rb-binding and the subsequent release of otherwise inactive E2F transcription factors. The counteractive response of affected cells is the expression of p53, which normally results in apoptosis. However, adenoviral E1B-55K can inactivate p53, resulting in progression to S phase and efficient viral replication (Yew and Berk, 1992). The adenovirus mutant *dl*1520 is therefore "theoretically" deprived of completing its life cycle in cells with functional p53. Nevertheless, *dl*1520 will be able to do so if p53 is not functional, such as in the case of most human cancers (Greenblatt et al., 1994). That this approach was a simplification became evident when an essential role for E1B-55K for later events in the adenoviral life cycle was revealed (Harada and Berk, 1999; Holm et al., 2002). Overall, the replication efficiency of this E1B-55K mutant seldom reaches the level of wild-type adenoviruses (Howe et al., 2000). Other E1B-modified oncolytic adenoviruses were generated by deletion of both E1B-55K and E1B-19K, which additionally targets Rb negative cancers that inhibit apoptosis (Duque et al., 1999). Deletions in the E1A region render adenovirus replication-specific for cells with inactive Rb. Two conserved regions in E1A, constant region 1 (CR1) and 2 (CR2), are essential to bind Rb. With respect to oncolytic adenoviruses, CR1-deletion mutants are barely selective and have an undesired attenuation in tumor cells (Heise and Kirn, 2000). On the contrary, a single 24bp deletion in CR2 preserves potent oncolytic activity in Rb negative tumor cells and renders these mutants replication deficient in normal cells (Fueyo et al., 2000; Heise and Kirn, 2000). In short, although mutations introduced into E1A and/or E1B genes increased the tumor specificity of viral replication, in some cases they also decreased efficiency of viral replication. Adenoviral vectors with deletions in E1A and E1B have the ability to replicate their DNA in a variety of human tumor cell lines but not in primary human cells *in vitro* or in livers of mice and nonhuman primates after systemic application (Steinwaerder et al., 2001).

Transcriptional targeting has also been used in order to restrict viral E1A expression and subsequent replication of adenoviruses to tumor cells. Utilized promoters include the α–fetoptotein promoter (Hallenbeck et al., 1999), the prostate-specific antigen (PSA) promoter (Rodriguez et al., 1997), and a mucin-like DF3 antigen (Muc1/DF3) promoter (Kurihara et al., 2000). These tissue specific adenoviruses demonstrated oncolytic abilities in corresponding cancer cell lines and in *in vivo*-models. However, tissue-specific expression cannot exclude transcriptional leakiness due to viral enhancer sequences. This might lead to E1A expression in other tissues, when these vectors are applied in high multiplicities of infection (MOIs) (Jounaidi et al., 2007). Increased tumor specificity for oncolytic adenoviruses can be achieved by combination of several earlier mentioned targeting strategies, e.g. employing transcriptional and transductional targeting, as well as cancer-selective replication. An example is Ad5/35Δ24Ki/COX, which was generated by Oliver Wildner's laboratory (Hoffmann et al., 2008). This Ad5-based vector is re-targeted to CD46 via fiber swapping to serotype 35. It also contains the 24bp deletion in CR2 of E1A and uses the Ki-67 and COX-2 promoters to express viral E1 and E4, respectively. The Ki-67 promoter is preferentially active in proliferating cells (Hoffmann and Wildner, 2006) and the activity of the COX-2 promoter is upregulated in a variety of tumor cells including gastrointestinal cancer (see 1.4.2.2.) and melanoma (Nettelbeck et al., 2003).

INTRODUCTION

*1.4.2.4. Arming of oncolytic adenoviruses*

The efficacy of oncolytic adenoviruses can be increased by insertion of a transgene cassette into the viral backbone. The expression of additional transgenes usually has the goal to eliminate surrounding cells and therefore enhance viral spread in solid tumors (Alemany, 2007). Examples for such approaches are the expression of the adenovirus death protein or p53 in order to enhance viral oncolysis (Doronin et al., 2000; Sauthoff et al., 2002). Furthermore, the expression of a mutated form of IκB was used to sensitize cells to TNF-mediated apoptosis (Mi et al., 2001). Oncolytic adenoviruses are also proposed as adjuvants for tumor immunotherapy by debulking the tumor in order to release tumor antigens and additionally induce inflammatory signals. In this context, the adenovirus-mediated expression of several interleukins has shown enhanced anti-tumor efficacy (Bortolanza et al., 2009; Lee et al., 2006; Post et al., 2007).

*1.4.2.5. Oncolytic vector Ad5/35.IR-E1A/TRAIL*

Ad5/35.IR-E1A/TRAIL is an oncolytic vector based on a 'first-generation' Ad5 that is devoid for E1A and E1B. It carries the RSV promoter upstream of a bicistronic expression cassette coding for adenoviral E1A (from Ad5), an internal ribosomal entry sequence (IRES), and tumor necrosis factor-related apoptosis-inducing ligand (TRAIL, also known as Apo2). The expression cassette is placed in an inverted 3'-5' position, which inhibits transgene expression, and is flanked by inverted repeats (IR) (Fig. 1.8). Further tumor specificity is achieved by incorporation of the short-shafted adenovirus serotype 35 fiber (Ad5/35), which targets this vector to CD46 (as mentioned in 1.4.2.1.).

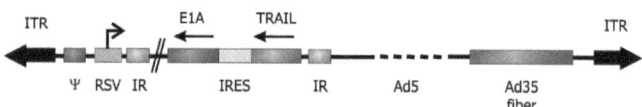

**Figure 1.8: Genome of Ad5/35.IR-E1A/TRAIL.** The vector contains the Ad5 genome deleted for E1A and E1B. The Ad5 fiber gene is replaced by the Ad35 fiber. The RSV promoter is placed upstream of an inverted transgene cassette, which is flanked by inverted repeats (IR). As IRs function two rabbit globin introns that are spliced out upon transcription and do not contain any transcription stop sites. The bidirectional SV40 polyadenylation signal ( **//** ) is used to terminate transcription. Adenovirus 5 E1A and TRAIL are in a bicistronic transgene cassette, connected by an IRES that confers simultaneous expression. Ψ = Adenovirus packaging signal.

In tumor cells that support low level viral replication (Engelhardt et al., 1994; Steinwaerder et al., 2001), homologous recombination between IRs in adjacent adenoviral genomes leads to placement of the RSV promoter in front of the 5' end of the expression cassette. Expression of E1A further amplifies vector DNA replication in a positive feedback manner, which increases the frequency of recombination and subsequently the expression of transgenes, and furthermore, results in production of progeny virus (Bernt et al., 2002). Both transgenes, E1A (Rao et al., 1992) and TRAIL (Pitti et al., 1996; Wiley et al., 1995), have proapoptotic properties and the potential to cause efficient tumor cell death. In normal cells that do not support viral replication, recombination and subsequent transgene expression are inhibited. Ad5/35.IR-E1A/TRAIL has been proofed efficient in eliminating human cell lines derived from colorectal, liver, lung, and prostate cancer. Additionally, in a model for metastatic colon cancer, tail vain infusion of Ad5/35.IR-E1A/TRAIL resulted in elimination of pre-established liver metastases in immunocompromised mice (Sova et al., 2004). The oncolytic performance of Ad5/35.IR-E1A/TRAIL was also studied in a glioblastoma model, where intratumoral injection could significantly delay tumor growth of subcutaneous tumors in NOD/SCID mice (Wohlfahrt et al., 2007). However, total elimination of tumors was not achieved in these studies.

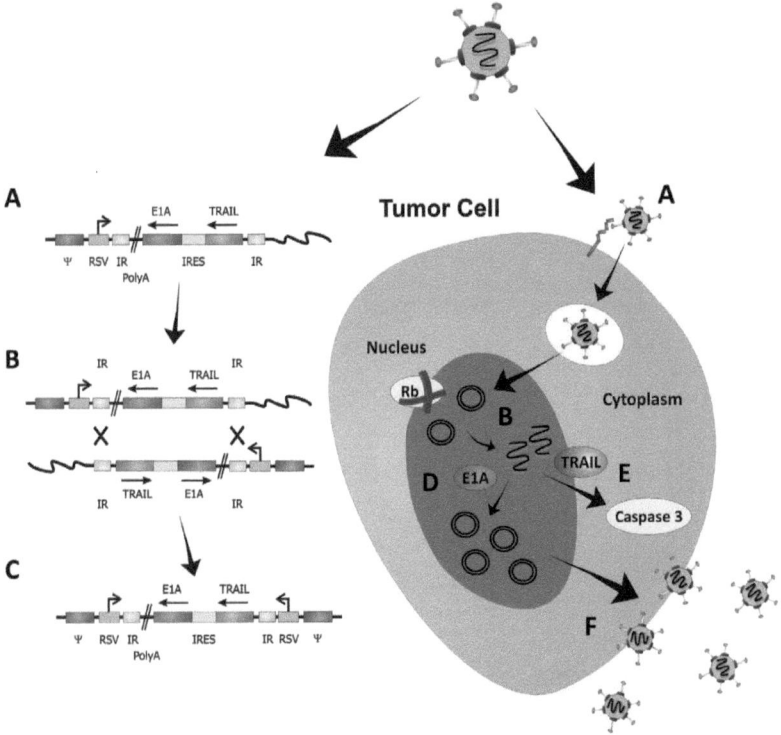

**Figure 1.9: Life cycle of oncolytic vector Ad5/35.IR.E1A/TRAIL. A)** Viral particles attach to CD46 and get internalized into tumor cells. **B)** After localization of the viral genome to the nucleus, low-level viral DNA replication occurs in Rb-deficient cells. This enables homologous recombination between IRs in adjacent viral genomes that leads to the generation of a rearranged genome carrying two inverted copies of the left half of the parental genome (including the promoter flanking the transgene). **C)** In the recombinant genome product, the transgenes are in correct (5'-3') orientation, allowing for expression of TRAIL and E1A from the RSV promoter. **D)** The expression of E1A enhances viral replication of the parental vectors, which in turn leads to more recombinant events and a further increase in E1A expression. **E)** The expression of TRAIL leads to caspase 3-induced apoptosis. **F)** Cell death is additionally induced as a consequence of the replication process of adenoviruses, which results in the release of progeny virus after cell lysis.

## 1.4.3. Clinical trials with oncolytic adenoviruses

Clinical trials with oncolytic adenoviruses were initiated in 1996 with the direct injection of dl1520 (see 1.4.2.3.) into head and neck cancers (Ganly et al., 2000). During the following years this vector was used under the name ONYX-015 in a total of 18 clinical trials (Phase I and II) with almost 300 treated patients (Alemany, 2007; Yu and Fang, 2007). In the initial trial about 14% of patients showed tumor regression rates of >50%, when ONYX-015 was used as a single agent (Ganly et al., 2000). Better results were obtained when oncolytic virus-administration was combined with chemotherapy in a Phase II trial (Khuri et al., 2000). Here, in about 65% of treated patients with head and neck tumors objective responses were seen, when compared to 30% response rates in patients that only received chemotherapy (cisplatin or 5-fluorouracil). However, long-time survival of patients with responses to adenovirus treatment did not significantly increase. The first oncolytic adenovirus that completed a Phase III trial and gained marketing approval for cancer treatment in China is H101 (Yu and Fang, 2007). This vector is highly similar to dl1520, but also deleted for E3. H101 is intended for intratumoral injection of head and neck cancers or other accessible solid tumors in combination with chemotherapy. Similar to studies with ONYX-015, in the Phase III trial 79% of patients responded to such type of treatment, whereas only 30-40% responded to chemotherapy alone (Yu and Fang, 2007). Table 1.2 shows a selection of the described clinical trials and others using oncolytic adenoviruses.

| Adenovirus | Tumor-specificity | Phase | Cancer | Administration | Response in patients | Reference |
|---|---|---|---|---|---|---|
| dl1520 (ONYX-015) | E1B-55 kDa-deletion | I | Head & Neck | i.t. | 3/22 | (Ganly et al., 2000) |
| dl1520 (ONYX-015) | E1B-55 kDa-deletion | II | Head & Neck | i.t. | 4/30 | (Nemunaitis et al., 2001b) |
| dl1520 (ONYX-015) | E1B-55 kDa-deletion | II | Head & Neck | i.t. + chemotherapy | 19/30 | (Khuri et al., 2000) |
| dl1520 (ONYX-015) | E1B-55 kDa-deletion | I | Pancreatic | i.t. | 0/23 | (Mulvihill et al., 2001) |
| dl1520 (ONYX-015) | E1B-55 kDa-deletion | I | Ovarian | i.p. | 0/16 | (Vasey et al., 2002) |
| dl1520 (ONYX-015) | E1B-55 kDa-deletion | I | Metastaic lung | i.v. | 0/10 | (Nemunaitis et al., 2001a) |
| dl1520 (ONYX-015) | E1B-55 kDa-deletion | II | Colorectal | i.v. | 0/18 | (Hamid et al., 2003) |
| dl1520 (ONYX-015) | E1B-55 kDa-deletion | I | Glioma | i.t. | 3/24 | (Chiocca et al., 2004) |
| CV787 (CG7870) | E1A under probasin and E1B under PSA promoter | I-II | Prostate | i.v. | 0/23 | (Small et al., 2006) |
| CV706 | E1A expression under PSA promoter | II | Prostate | i.t. | 5/20 | (DeWeese et al., 2001) |
| H101 | E1B-55 kDa-deletion | I-II | Multiple | i.t. | 3/15, 14/46 | (Yu and Fang, 2007) |
| H101 | E1B-55 kDa-deletion | III | Head & Neck | i.t. + chemotherapy | 41/52 | Yu and Fang, 2007) |

**Table 1.2: Selected clinical trials with oncolytic adenoviruses.** Listed are used viruses, their cancer-specificity, clinical trial phases they passed, the cancer type they were used for, their route of administration, and the number of patients who objectively responded to the treatment. i.t.= intratumoral, i.p.=intraperitoneally, i.v.=intravenously, PSA=prostate specific antigen.

Overall, virus administration of up to $1 \times 10^{13}$ viral particles (about $5 \times 10^{11}$ pfu) was generally well tolerated. Common symptoms after viral injections were flu-like symptoms, fever, pain in the injection site, and nausea. Notably, no tumor responses could be observed after systemic or intraperitoneal application of oncolytic adenoviruses.

**1.4.4. Limitations for oncolytic adenoviruses**

Although oncolytic adenoviruses based on serotype 5 have proved efficient in animal models and safe in patients, they have fallen short of their expected therapeutic value as monotherapies in clinical trials (Hermiston, 2006). The lack of CAR expression on tumor cells *in situ* is likely one of the reasons and the role of this will become clearer when data from ongoing clinical trials with capsid-modified adenoviruses will be publicly available. However, there is a series of other obstacles that limit the efficacy of oncolytic adenoviral vectors. These include innate host defense mechanisms that recognize adenoviruses as a pathogen are aimed towards eliminating it.

*1.4.4.1. Innate responses to adenovirus*

Systemic delivery of adenoviruses has the potential to target disseminated tumors. However, a number of soluble and cellular blood components, including the complement factors (Jiang et al., 2004; Kiang et al., 2006), blood coagulation factors (Kalyuzhniy et al., 2008; Waddington et al., 2008), pre-existing neutralizing antibodies (Moskalenko et al., 2000; Zhi et al., 2006), erythrocytes and platelets (Stone et al., 2007) interact with adenovirus particles, resulting in virus degradation or sequestration by the reticulo-endothelial system mainly of the liver. Specifically, for a number of adenovirus serotypes, including Ad5, tumor-targeting is critically limited due to the high-affinity interaction between blood coagulation factor X (FX) and viral hexon (Liu et al., 2009), which triggers HSPG-mediated adenovirus uptake into hepatocytes (Kalyuzhniy et al., 2008; Vigant et al., 2008; Waddington et al., 2008). Furthermore, it is thought that unspecific sequestration of Ad5 particles by the reticular-endothelial system is the major factor that limits tumor targeting in mice. In humans, additional obstacles such as Ad5 binding to erythrocytes via CAR or opsonization of anti-Ad5 immune complexes most likely critically affect the efficacy of Ad5-based oncolytic viruses. Recent reports also revealed intracellular innate response mechanisms. One of these mechanisms involves direct binding of $\alpha$-defensins to the capsid of Ad5 and Ad35, which prevents conformational changes required for endosomal escape after entry into host cells (Smith and Nemerow, 2008). Defensins are small, secreted, cationic peptides that are directly toxic to microbes and have signaling functions in the inflammatory response. In humans they are expressed by neutrophils, and epithelial cells of the small intestine, female genitourinary tract, and airway (Ganz, 2003). Another intracellular defense mechanism is based on cathelicidin (LL–37/hCAP–18), a protein expressed by epithelial cells with similar functions to defensins, that has been shown to significantly reduce infectious titers of Ad19 on A549 cells *in vitro* (Gordon et al., 2005). At this point, it remains to be documented that these intracellular defense mechanism are active in tumor cells. Finally, the extremely genetic and phenotypic plasticity and mutation rate of tumor cells, often associated with a so called "mutator phenotype" (Loeb et al., 2008) has to be included in the list of "innate" defense mechanism that potentially limit the efficacy of virotherapy or lead to emergence of cancer cell subsets that can escape infection with oncolytic adenoviruses.

*1.4.4.2. Tumor stroma*

For solid tumors, that reach a minimal size of 1 to 2 mm, the development of tumor stroma becomes essential (Folkman and Shing, 1992). Tumor stroma is composed of stroma cells and a complex structured extracellular matrix (ECM), containing collagen, laminin, glycosaminoglycans, hyaluronic acid, and proteoglycans. Stroma cells include fibroblasts, endothelial cells, pericytes and tumor-infiltrating hematopoietic cells, predominantly derived from myeloid lineage progenitor cells located in the bone marrow. ECM components form extensive networks and connect to tumor cells via cellular integrins. It is thought that ECM proteins are largely a product of tumor stroma cells. Compared to normal tissue, the ECM of solid tumors is subjected to constant remodeling processes, which can induce migration and invasion of cancer cells (Cheng et al., 2007a). Tumor stroma attenuates the efficacy of anti-tumor immune cells, oncolytic adenoviruses (Li et al., 2004), antibodies (Jain, 1990), immunotoxins (Wenning and Murphy, 1999), IFNs (Gorlach et al., 1994), or complement (Bjorge et al., 1997). Overall it is thought that the large size of many blood-borne therapeutics (such as proteins, liposomes, or nanoparticles) impairs transport through the tumor extracellular matrix and thus limits their therapeutic effectiveness. A number of approaches have been used to transiently degrade the ECM in order to increase adenovirus transduction and intratumoral dissemination. Intratumoral injection of collagenase has been shown to remove diffusive hindrance to the penetration of therapeutic molecules within tumors (Brown et al., 2003). Also, direct administration of collagenase/dispase or trypsin into glioma xenografts has been shown to enhance the extent of infection of a non-replicating adenoviral vector expressing a reporter gene (Kuriyama et al., 2000). Two recent studies demonstrated that virus-mediated intratumoral expression of MMP1 and /or MMP8 can degrade collagen, reduces the levels of sulfated proteoglycans, and improves the spread of oncolytic adenovirus (Cheng et al., 2007b) or herpes simplex virus (Mok et al., 2007). Another protein that was used to degrade tumor stroma proteins is relaxin (RLX), a peptide hormone that, during pregnancy, is involved in softening the uterine cervix, vagina, and interpubic ligaments in preparation for parturition (Sherwood, 2004). Recently, it was showed that inducible, intratumoral expression of relaxin, either by transplanting tumor cells that contained the relaxin gene or by transplantation of mouse hematopoietic stem cells transduced with a relaxin-expressing lentivirus vector delays tumor-growth in an immune-competent mouse breast cancer model (Li et al., 2009). The antitumor effect of relaxin was mediated through degradation of tumor stroma, which provided increased access of immune cells. In this study it was also demonstrated that intratumoral relaxin expression improves tumor transduction with an adenoviral vector.

## 2. Aim of the study

This study was aimed towards the identification of cellular mechanisms that render epithelial ovarian cancer cells resistant to adenovirus-mediated oncolysis. To achieve this goal, the first task was the establishment of *in vitro* cultures that mirror the resistance that is generally observed in solid tumors. The employment of an *in vitro* system was a critical step in order to use established techniques that monitor and compare virus-cell interactions (e.g. internalization process, trafficking) on cells either resistant or susceptible to viral oncolysis. It was hypothesized that primary cultures better reflect the diversity of cancer cells that is observed in solid tumors than established cell lines and therefore would contain tumor cell subsets that could escape viral oncolysis.

Once primary cell cultures that are resistant to viral oncolysis were identified, it had to be investigated on which level the resistance to adenovirus-induced cell death is conferred. In order to identify alterations in pathways that are involved in these processes, the gene expression profile of cells either resistant or susceptible to adenoviral oncolysis was subjected to genome-wide DNA expression array analysis. Considering the complexity of the used oncolytic adenoviruses, efficient oncolysis could be affected on at least five different levels: *i)* the viral attachment to cellular receptors, *ii)* virus internalization and subsequent trafficking of viral particles to the nucleus, *iii)* the replication of the viral genome, *iv)* the expression of adenoviral genes, including the transgenes E1A and TRAIL, or *v)* the release of progeny virus.

The identified mechanisms that render *in vitro* cultures resistant to adenoviral oncolysis will then have to be tested in ovarian cancer xenografts and tumors *in situ*. Cell lines derived from other epithelial cancers and different adenoviral vectors targeting CAR and CD46 will be employed to generalize the observed resistance mechanisms. In order to overcome the resistance to adenoviral infection and/or oncolysis, affected pathways will be manipulated. Finally, it will be explored whether the diversity among various adenovirus serotypes offers candidates that are better suited for the generation of future oncolytic vectors.

# 3. Material and methods

## 3.1. Material

### 3.1.1. Antibodies

The following antibodies were used for immunofluorescence experiments (Table 3.1), flow cytometry (Table 3.2) and western blotting (Table 3.3). Listed are the antibody target proteins (including direct fluorochrome conjugates), the host where the antibody was raised in, the dilution factor used in the experiments, and the vendor where the antibody was purchased from using the appropriate catalog number. If not stated otherwise, antibodies were directed against human antigens.

| Antibody target | Host | Dilution | Vendor | Catalog number |
|---|---|---|---|---|
| β-catenin-FITC | mouse | 1:50 | BD Transduction Lab. | 610156 |
| CAR | mouse-hybridoma | 1:5 | ATCC | RmcB |
| CD31 (mouse) | rat | 1:100 | BD Pharmingen | 553708 |
| CD46 | mouse | 1:50 | Fitzgerald | clone J4.48 |
| Cingulin | rabbit | 1:100 | Zymed | 36-4401 |
| Claudin1 | rabbit | 1:50 | Zymed | 51-9000 |
| Claudin2 | mouse | 1:50 | Zymed | 32-5600 |
| Claudin3 | rabbit | 1:100 | abcam | ab15102 |
| Claudin4 | rabbit | 1:100 | abcam | ab15104 |
| Claudin7 | rabbit | 1:100 | abcam | ab27487 |
| Collagen IV | goat | 1:400 | Southern Biotech | 1460-01 |
| E-cadherin | mouse | 1:200 | Sigma | U 3254 |
| E-cadherin-FITC | mouse | 1:100 | BD Biosciences | 612130 |
| EpCAM | mouse | 1:100 | Biomeda | FM010 |
| Fibronectin | rabbit | 1:100 | abcam | ab2413 |
| Hexon (adenovirus) | mouse | 1:200 | Chemicon | MAB8052 |
| Hexon (adenovirus)-Cy3 | mouse | 1:20 | Chemicon/Amersham | MAB8053/PA23000 |
| Integrin-$\alpha_V$ | mouse-hybridoma | 1:40 | ATCC | HB-8448 |
| Laminin | rabbit | 1:400 | Dako | M0638 |
| N-cadherin | rabbit | 1:150 | GeneTex | GTX 12221 |
| Netrin4 | goat | 1:20 | R&D Systems | AF1254 |
| Occludin-FITC | mouse | 1:100 | Zymed | 33-1511 |
| p120-catenin | rabbit | 1:100 | abcam | ab32095 |
| Phalloidin-AF488 | mushroom | 1:100 | Invitrogen/ Mol. Probes | A12379 |
| Phalloidin-TRITC | mushroom | 1:100 | Chemicon | FAK100 |
| Vimentin | goat | 1:100 | Sigma | HPA001762 |
| Vinculin | mouse | 1:100 | Chemicon | FAK100 |
| Anti-rabbit-AF488 | goat | 1:200 | Invitrogen/ Mol. Probes | A-11008 |
| Anti-rabbit-AF568 | goat | 1:200 | Invitrogen/ Mol. Probes | A-11011 |
| Anti-mouse-AF488 | goat | 1:200 | Invitrogen/ Mol. Probes | A-11001 |
| Anti-mouse-AF568 | goat | 1:200 | Invitrogen/ Mol. Probes | A-11004 |
| Anti-goat-AF568 | donkey | 1:200 | Invitrogen/ Mol. Probes | A-11057 |
| Anti-rat-AF488 | goat | 1:200 | Invitrogen/ Mol. Probes | A-11006 |
| Anti-mouse-AF405 | goat | 1:200 | Invitrogen/ Mol. Probes | A31553 |

Table 3.1: Antibodies for immunofluorescence experiments used in this study. AF=Alexa Fluor, FITC=fluorescein isothiocyanate, TRITC=tetramethylrhodamine isothiocyanate, Cy3=Cyanine Dye 3

| Antibody target | Host | Dilution | Vendor | Catalog number |
|---|---|---|---|---|
| EpCAM-FITC | mouse | 1:10 | Stem Cell Technologies | 10109 |
| CD44-PE | mouse | 1:10 | BD Pharmingen | 555479 |
| E-cadherin-AF488 | mouse | 1:20 | Biolegend | 324110 |
| Vimentin-PE | mouse | 1:10 | abcam | ab49918 |
| p120-catenin | rabbit | 1:33 | abcam | ab32095 |
| Anti-rabbit-PE | goat | 1:100 | Invitrogen/ Mol. Probes | P-2771MP |

**Table 3.2: Antibodies for flow cytometry used in this study.** AF=Alexa Fluor, FITC=fluorescein isothiocyanate, PE=R-Phycoerythrin

| Antibody target | Host | Dilution | Vendor | Catalog number |
|---|---|---|---|---|
| FAK | rabbit | 1:1000 | Cell Signaling | 3285 |
| PI3K p85 | rabbit | 1:1000 | Cell Signaling | 4292 |
| Phospho-PI3K p85 /p55 | rabbit | 1:1000 | Cell Signaling | 4228S |
| Rho | mouse | 1:1000 | Cell Biolabs | 240302 |
| Rac1 | mouse | 1:1000 | Cell Biolabs | 240106 |
| Cdc42 | mouse | 1:1000 | Cell Biolabs | 240201 |
| Caldesmon | rabbit | 1:1000 | abcam | ab45691 |
| ROCK1 | rabbit | 1:1000 | Cell Signaling | 4035 |
| E-cadherin | rabbit | 1:1000 | Cell Signaling | 3195 |
| p120-catenin | rabbit | 1:1000 | abcam | ab32095 |
| GAPDH | mouse | 1:1000 | abcam | ab9484 |

**Table 3.3: Antibodies for western blotting used in this study.**

### 3.1.2. Inhibitors

The following inhibitors were used in the study. Listed are the inhibitor target, the used final concentration, the solution of the stock fluid (control), and the vendor from which inhibitors were purchased (including order number). The final concentrations were used in the indicated literature before.

| Inhibitor | Final concentration | Control | Vendor | Reference |
|---|---|---|---|---|
| Phosphoinositide 3-kinase (PI3K) inhibitor (Wortmannin) | 1 µM | DMSO (1:5000) | Calbiochem (#681675) | (Li et al., 1998a) |
| Rac 1 inhibitor (InSolution™ Rac1 Inhibitor) | 100 µM | dH$_2$O | Calbiochem (#553508) | (Gao et al., 2004) |
| Rho A,B,C inhibitor (*Clostridium botulinum* Exoenzyme C3) | 1 µg/ml | dH$_2$O | Calbiochem (#341208) | (Genth et al., 2003) |
| Rho kinase (ROCK) inhibitor (H-1152) | 100 µM | dH$_2$O | Calbiochem (#555550) | (Ikenoya et al., 2002) |
| Rho GTPase inhibitor (*Clostridium difficile* toxin B) | 100 ng/ml | dH$_2$O | Calbiochem (#616377) | (Li et al., 1998a) |

**Table 3.4: Inhibitors used in this study.**

### 3.1.3. Mouse strain

CB-17/lcrCrl-scid-bgBR (CB17-SCID-beige) mice were purchased from Charles River Laboratories. All experiments involving animals were conducted in accordance with the institutional guidelines set forth by the University of Washington. All mice were housed in specific pathogen-free facilities.

### 3.1.4. Adenoviruses

The following adenoviruses were used in this study. Listed are viruses, their deletions, the transgenes they carry, the promoters used for expression of transgenes, and their origin.

| Adenovirus | Deletions | Transgene/s | Transgene promoter/s | Origin (reference) |
|---|---|---|---|---|
| Ad5/35Δ24Ki/COX | E1A CR2 (Δ24), E3 | - | Ki-67 (E1 expression), COX-2 (E4 expression) | Oliver Wildner (Hoffmann et al., 2008) |
| Ad5/35.IR-E1A/TRAIL | E1, E3 | E1A, TRAIL | RSV | lab-internal vector (Sova et al., 2004) |
| Ad5.IR-E1A/TRAIL | E1, E3 | E1A, TRAIL | RSV | lab-internal vector (Sova et al., 2004) |
| Ad5/35.GFP | E1, E3 | GFP | CMV | lab-internal vector (Shayakhmetov et al., 2000) |
| Ad5.GFP | E1, E3 | GFP | CMV | lab-internal vector (Shayakhmetov et al., 2000) |
| Ad5/35.β-gal | E1, E3 | β-gal | RSV | lab-internal vector (Shayakhmetov et al., 2004) |
| Ad3 | - | - | - | GB strain (ATCC) |
| Ad5 | - | - | - | Reference strain (ATCC) |
| Ad7p | - | - | - | Gomen strain (ATCC) |
| Ad11p | - | - | - | Slobitky strain (ATCC) |
| Ad14 | - | - | - | DeWit strain (ATCC) |
| Ad35 | - | - | - | Holden strain (ATCC) |

**Table 3.5: Adenoviruses used in this study.** TRAIL=tumor necrosis factor-related apoptosis-inducing ligand, GFP=green fluorescent protein, β-gal=beta-galactosidase, COX-2=Cyclooxygenase-2, RSV=rous sarcoma virus, CMV=cytomegalovirus

### 3.1.5. Oligonucleotides

All oligonucleotides were purchased as lyophilized, salt-free stocks from Operon. The following tables list oligonucleotides used for the detection of adenoviral gene expression and viral genomes (Table 3.7 and 3.9), or validation of DNA expression arrays by qRT-PCR (Table 3.8 and 3.9). The official gene symbol (by Human Genome Organization (HUGO) Gene Nomenclature Committee) was used for oligonucleotide (primer) names.

| Primer | Direction | Sequence |
|---|---|---|
| Ad5 hexon | Forward | 5' TACTGCGTACTCGTACAAGG 3' |
| Ad5 hexon | Reverse | 5' AGAGCAGTAGCAGCTTCATC 3' |
| Ad5 E2A DBP | Forward | 5' AAGCGGATGAGGCGGCGTAT 3' |
| Ad5 E2A DBP | Reverse | 5' GTAGCGCCACATCTTCTCTT 3' |
| Ad5 pIX | Forward | 5' GGTGCGTCAGAATGTGATGG 3' |
| Ad5 pIX | Reverse | 5' GAGCCGTCAACTTGTCATCG 3' |
| Ad5 E1A | Forward | 5' TCTGCCACGGAGGTGTTATT 3' |
| Ad5 E1A | Reverse | 5' TTCCTGCACCGCCAACATTA 3' |
| TRAIL | Forward | 5' GGACAGACCTGCGTGCTGAT 3' |
| TRAIL | Reverse | 5' CGGAGTTGCCACTTGACTTG 3' |

**Table 3.6: Oligonucleotides for adenoviral gene expression and viral genome detection.**

# Material and Methods

| Primer | Direction | Sequence |
|---|---|---|
| MYLK | Forward | 5' GAGCTGGAGAACTCCATGTA 3' |
| MYLK | Reverse | 5' GACGCAAGTCTGAGTGCATT 3' |
| RRAS2 | Forward | 5' CACGGCAGCTTAAGGTAACA 3' |
| RRAS2 | Reverse | 5' TCCTGGCCGTTGGTAGCTAA 3' |
| RHOA | Forward | 5' ACTGTCATCCTCAAAGAAAG 3' |
| RHOA | Reverse | 5' CTAATTGCCTCAGGCGTGAA 3' |
| ACTN1 | Forward | 5' GACAGCCGACACAGATACAG 3' |
| ACTN1 | Reverse | 5' AGGTCACTCTCGCCGTACAG 3' |
| CLDN4 | Forward | 5' GGAGTATGGCTGAGGCCTTG 3' |
| CLDN4 | Reverse | 5' CAGGCTCATTAGTGTCCTTG 3' |
| CLDN23 | Forward | 5' GCCAGCAGCTTAATGGATTT 3' |
| CLDN23 | Reverse | 5' GTCCTGGATGAGGCGTCCAA 3' |
| CLDN7 | Forward | 5' ATGGACTGCGTCACGCAGAG 3' |
| CLDN7 | Reverse | 5' CCACAGCGCGTGCACTTCAT 3 |
| CLDN2 | Forward | 5' TCAGCGTCACCTCCTTCATT 3' |
| CLDN2 | Reverse | 5' CCTGGCACTGTTACAGATAG 3' |
| NTN4 | Forward | 5' AAGCAGCAGCACTGCCACTA 3' |
| NTN4 | Reverse | 5' CATCATGGCTGTCATCTTGG 3' |
| CDH1 | Forward | 5' TGGCAGGAGAGCTTGTCATT 3' |
| CDH1 | Reverse | 5' TGTTCAGCTCAGCCAGCATT 3' |
| CD99 | Forward | 5' CGTGGCTGGAGCCATCTCTA 3' |
| CD99 | Reverse | 5' GAGCCTCAGGCAGCTGTTCT 3' |
| RAP1B | Forward | 5' TGGCTCAGGAGGCGTTGGAA 3' |
| RAP1B | Reverse | 5' ATGTGGACTGTGCTGTGATG 3' |
| MYH10 | Forward | 5' CAGCCATGGATCATGTAGAC 3' |
| MYH10 | Reverse | 5' GGAACAAGCCTCGGACACAA 3' |
| VCAM1 | Forward | 5' CTGAGCTTCTCGTGCTCTAT 3' |
| VCAM1 | Reverse | 5' CTGCCTCTCAGCTCATTGTT 3' |
| CFL1 | Forward | 5' CAACGCCAGAGGAGGTGAAG 3' |
| CFL1 | Reverse | 5' TGCTCTCCTTGGTCTCATAG 3' |
| Cdc42 | Forward | 5' TGAGTTGCCTGATGCTCAGA 3' |
| Cdc42 | Reverse | 5' CCCCATTACACTCTACTAGG 3' |
| Rac1 | Forward | 5' CACTGTCTTGCCAGATTACC 3' |
| Rac1 | Reverse | 5' TCTGTTGTAGTGGCTGAAGG 3' |
| ROCK2 | Forward | 5' CCAACTGTGAGGCTTGTATG 3' |
| ROCK2 | Reverse | 5' CCACTTCTGCTGCTCTTCTG 3' |
| ROCK1 | Forward | 5' AGGCCTTCTGGAAGAACAGT 3' |
| ROCK1 | Reverse | 5' GGCAGCCTTAAGATTACTGA 3' |
| OCLN | Forward | 5' AGTACATGGCTGCTGCTGAT 3' |
| OCLN | Reverse | 5' ACAACTTGGCATCAGCCTTC 3' |
| CTNNB1 | Forward | 5' TCCGAATGTCTGAGGACAAG 3' |
| CTNNB1 | Reverse | 5' CAAGGCATCCTGGCCATATC 3' |
| JNK | Forward | 5' AACTGCAACCAACAGTAAGG 3' |
| JNK | Reverse | 5' GATGTACGGGTGTTGGAGAG 3' |
| DIAPH1 | Forward | 5' CCACAGGAAGCTGCAATTCT 3' |
| DIAPH1 | Reverse | 5' TACACAAGCGCCAGCAACCA 3' |
| CLDN1 | Forward | 5' GCCACTGTGTCTTATGAGGA 3' |
| CLDN1 | Reverse | 5' CCAGGTGTGGTAGAAGAATA 3' |
| CLDN3 | Forward | 5' CCATCCAGCGTGCAGCCTTG 3' |
| CLDN3 | Reverse | 5' GGTCAAGTATTGGCGGTCAC 3' |

Table 3.7: Oligonucleotides for DNA expression array validationq by qRT-PCR.

| Primer | Direction | Sequence |
|---|---|---|
| GAPDH | Forward | 5' AGGTGGTCTCCTCTGACTTC 3' |
| GAPDH | Reverse | 5' CTCTTCCTCTTGTGCTCTTG 3' |
| HPRT1 | Forward | 5' AGTTCTGTGGCCATCTGCTT 3' |
| HPRT1 | Reverse | 5' GCCCAAAGGGAACTGATAGTC 3' |
| CYPA | Forward | 5' CGGGTCCTGGCATCTTGT 3' |
| CYPA | Reverse | 5' GCAGATGAAAAACTGGGAACCA 3' |

Table 3.8: Oligonucleotides for housekeeping genes.

### 3.1.6. Cultured cells and culture media

The listed cells and culture media were used throughout this thesis. For cell passaging, cells were detached from tissue culture plates (BD Falcon) with TrypleExpress (Gibco) and then washed with RPMI 1640 supplemented with 10% FBS. To determine cell numbers, cultures were counted using a hemacytometer (improved Neubauer, Fisher Scientific). 1% Penicillin/Streptomycin (Gibco) was added to all media. Primary ovarian cancer cells were propagated in a 1:2 ratio and cultured in MEGM (MEBM containing 3mg/L hEGF, 5mg/L insulin, 5mg/L hydrocortisone, 26mg/L bovine pituitary extract, 25mg/L amphotericin B) in the additional presence of 10µg/ml Ciprofloxacin (CellGro) and 5mg/L Plasmocin (InvivoGen). All other used supplements are listed below. All cells were cultured at 37°C, 5% $CO_2$, and 95% humidity in cell culture incubators (Thermo Scientific). Cells were frozen in cryo tubes (Greiner) in 50% FBS, 40% of indicated medium (Table 3.9) and 10% dimethyl sulfoxide using a cell freezer (Nalgene) that contains isopropanol (Fisher Scientific).

| Cell type | Source | Description | Medium |
|---|---|---|---|
| Primary ovarian cancer cultures (including ovc316) | ovarian cancer patients, Swedish Hospital, Seattle, WA, USA | human epithelial ovarian cancer, different histology | MEGM (Lonza), 1-2% FBS (Gibco) |
| ovc316m | ovarian cancer patient, Swedish Hospital, Seattle, WA, USA | stage IV human serous ovarian carcinoma, isolated from mouse xenograft | MEGM (Lonza), 10% FBS (Gibco) |
| HEK-293 | Microbix, Toronto, Canada (Graham et al., 1977) | human embryo kidney cells, transformed by adenovirus serotype 5 E1A | DMEM (Gibco), 10% FBS (Gibco), 2 mmol/L L-glutamine (Gibco) |
| AE25 | Kovesdi, I. (Bruder et al., 2000) | human lung adenocarcinoma epithelial cells, adenoviral E1-complementing cell line derived from A549 cells | DMEM (Gibco), 10% FBS (Gibco), 2 mmol/L L-glutamine (Gibco) |
| SKOV3-ip1 | David T. Curiel (Kanerva et al., 2002) | human ovarian adenocarcinoma, tumorigenic subclone of SKOV3 | MEGM (Lonza), 10% FBS (Gibco) |
| HT-29 | American Type Culture Collection (ATCC), HTB-38™ (Fogh et al., 1977) | human colorectal adenocarcinoma | McCoy's 5A (Gibco), 1.5 mM L-glutamine (Gibco), 2.2 g/L sodium bicarbonate (Gibco), 10% FBS (Gibco) |
| HeLa | ATCC, CCL-2™ (Scherer et al., 1953) | human cervical adenocarcinoma | DMEM (Gibco), 10% FBS (Gibco) |

Table 3.9: Cells and their cell culture media. FBS=fetal bovine serum

### 3.1.7. Zonula occludens toxin

The Zonula occludens toxin (ZOT) was generously provided by Dr. Alessio Fasano (University of Maryland). Recombinant ZOT protein was produced in *E.coli* and purified using Nickel-nitrilotriacetic acid resin columns (Qiagen). Protein was stored in elution buffer (250mM imidazole, 8M urea, 0.1M sodium phosphate, 0.01M Tris-HCl, pH 6.3 [all Sigma]) at -80°C.

## 3.2. Methods

### 3.2.1. Primary tumor cell culture

Tumor tissue from biopsies was dissected into about 4mm pieces and digested with 1mg/ml collagenase/dispase (Roche) + 0.1mM $CaCl_2$ (Sigma) in RPMI 1640 (Gibco) for 2 hours at 37°C. This was followed by incubation with versene (Gibco) (1:1 vol/vol) for 1 hour. The protease digestion was stopped by the addition of FBS (Gibco) to a final concentration of 10%. Cells were passed through a 70µm cell strainer (BD Falcon) using the plunger of a 5ml syringe (Becton Dickenson). The cell strainer was washed with 25ml RPMI and the total flow-through was pelleted. The pellet was resuspended in 5ml RPMI/10%FBS with 1mg/ml DNaseI (Roche) and incubated for 30 min at 37°C. Cells were pelleted again and subjected to a ficoll gradient (10ml underlay, Stem Cell Technologies) centrifugation step if high amounts of erythrocytes were observed. The resulting cultures were kept in MEGM (Lonza) at low serum levels (1-2%) and in high cell densities in order to minimize fibroblast growth.

In order to establish single cell clones, primary cells were diluted to a concentration of 5 cells/ml and 100µl were seeded into 96well plates (BD Falcon) using a multi-channel pipette (Eppendorf). Cells were cultured in conditioned medium. Therefore, medium, used for 48 hours on primary cell culture, was filtered with a 0.22 µm filter (Becton Dickinson) and mixed with fresh medium in a ratio of 1:3 (used medium/fresh medium). Single cell plating was confirmed by light microscopy and wells containing more than one cell were excluded from further culturing. 50µl fresh medium was added every 5 days until clonal cultures were passaged.

### 3.2.2. Adenovirus propagation and preparation

Adenoviruses were propagated on HEK-293 cells in 150 mm dishes in a total volume of 20ml. For propagation of Ad5/35.IR-E1A/TRAIL and Ad5.IR-E1A/TRAIL the more apoptosis-resistant AE25 cell line was used. Cells were 90-100% confluent when infected. For initial infection, replication competent adenovirus (RCA)-free aliquots of virus-stocks were used in an approximate MOI of 10-25 pfu/cell. 5ml fresh medium was added the next day. When cells were rounded and started to de-attach (approximately 48 hours after infection), they were harvested in the culture medium by repeated pipetting. Cell-containing medium was transferred to a 50ml blue cap tubes (BD Falcon) and these then subjected to 4 cycles of freezing in liquid nitrogen and thawing at 37°C in a water bath. Tubes were centrifuged at 400xg (Beckman Coulter) and the supernatant was collected. Virus-containing supernatant was propagated on fresh HEK-293 or AE25 cells in a ratio of 1:3-1:4 until 30 150mm dishes were infected. Here, cells were collected when rounded, but before they started to de-attach (approximately 36 hours after infection). Cells were collected, pelleted (400xg) and then taken up in 1ml phosphate-buffered saline (PBS [Gibco BRL]) per plate. After 4 cycles of freezing and thawing, virus was isolated by ultracentrifugation. The first ultracentrifugation (2 hours, 14°C, 35,000 RPM, SW41 rotor [Beckman Coulter]) was performed in a Caesium chloride (CsCl) step gradient. The following CsCl concentrations were layered above each other in 12ml ultra-clear tubes (Beckman Coulter):

        *i)* 0.5 ml      1.50 g/cm$^3$ CsCl (45.41g CsCl + 54.49 ml H$_2$O)
        *ii)* 3.0 ml      1.35 g/cm$^3$ CsCl (35.18g CsCl + 64.82 ml H$_2$O)
        *iii)* 3.5 ml     1.25 g/cm$^3$ CsCl (26.99g CsCl + 73.01 ml H$_2$O).

5ml of viral supernatant were layered on top of the gradient and then tubes were centrifuged in a SW41 rotor for 1 hour at 35,000 RPM at 14°C (Beckman Coulter). Three clearly separated bands were obtained. Adenovirus appeared as a narrow, opaque white band in the lower 1/3 of the CsCl step gradient (Fig.3.1).

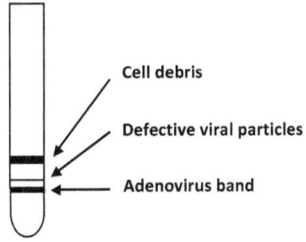

Fig. 3.1: Caesium chloride separation of adenovirus from defective particles and cell debris.

Adenovirus fractions of individual tubes were isolated and combined. 4ml were mixed with 8 ml 1.35 g/cm$^3$ CsCl in ultra-clear tubes and then centrifuged at 35,000 rpm overnight. The virus band was isolated from the bottom 1/4 of the tube and then dialyzed in a 50kDa cut-off dialyzing tube (Spectrum Laboratories) against 1,000 ml of 10 mM Tris pH 7.5, 10mM MgCl$_2$, 250mM NaCl and 10% glycerol overnight at 4°C with one change of dialyzing buffer. The virus was then collected and stored in 25µl or 50µl aliquots at -80°C.

### 3.2.3. Adenovirus titering by spectrophotometry

The adenovirus particle titer was determined on viral DNA. A 25µl aliquot from a fresh adenovirus stock was added to 475 µl TE buffer (10mM Tris pH 8.0, 1mM EDTA pH 8.0 [Sigma]) with 0.1% SDS (Sigma). The sample was thoroughly mixed using a vortexer (Baxter) for 5 min and then centrifuged at 14,000 RPM using an Eppendorf table centrifuge for 5 min. The optical density (OD) of supernatant was assessed at 260 nm in a spectrophotometer (Becton Dickenson). Viral particle titer/ml was calculated by multiplying the OD with $2 \times 10^{13}$ (Mittereder et al., 1996).

### 3.2.4. Adenovirus titering by plaque assay

HEK-293 cells were used to determine the plaque-forming unit titer of adenovirus stocks. Cells were plated in 6-well plates and infected when 90-100% confluent. An aliquot of the adenovirus stock was thawed on ice and then serial diluted in regular HEK-293 medium using the following dilutions: $10^{-2}$, $10^{-4}$, $10^{-6}$, $10^{-8}$, $10^{-9}$, $10^{-10}$. Cells were then infected in duplicates with 1ml virus dilution/well and incubated for 24 hours at 37°C. Medium was removed and cells then overlayed with 3ml of a warm (45°C) mix of 2xDMEM (Gibco) and 1.2% agarose (Sigma) (1:1 vol/vol) supplemented with 10% FBS. Cells were overlaid with 1 additional ml on days 4 and 9 after the first overlay. Plaques were counted on days 10 and 14 post infection and final titer was determined by multiplication of individual plaques with appropriate dilutions. The mean titer of duplicates was used.

### 3.2.5. Adenovirus-labeling with *methyl-$^3$H-thymidine*

$5 \times 10^7$ HEK-293 cells were grown in 175-cm$^2$ flasks with 15ml of DMEM+10% FCS and infected with Ad5/35 at an MOI of 50 pfu/cell. Twelve hours post infection, 1 mCi of [*methyl-$^3$H*]thymidine (Amersham) was added to the medium, and cells were further incubated at 37°C until complete cytopathic effect was observed. Then cells were harvested, pelleted, washed once with cold PBS (Gibco), and resuspended in 5 ml of PBS. Virus was released from the cells by four freeze-thaw cycles. Cell debris was removed by centrifugation, and viral material was subjected to ultracentrifugation in CsCl gradients and subsequent dialysis as described above. Titer was assessed spectrophotometrically (3.2.3.).

### 3.2.6. Virus Attachment/Uptake

$1 \times 10^5$ cells were seeded in 48well plates two days prior to infection. Confluent monolayers were washed with PBS once and then incubated with 200µl MEGM medium containing $^3$H-labeled adenovirus at an MOI of 8000 genomes per cell. After 30 min incubation at 4ºC, cells were washed twice with 0.5ml of ice-cold PBS. After the last wash, the supernatant was removed, and the cell-associated radioactivity was determined by a scintillation counter (LS 6500, Beckman Coulter). Virion specific radioactivity was determined using a dilution row of a known number of $^3$H-labeled viral particles ($1 \times 10^6$, $1 \times 10^5$, $1 \times 10^4$, 1000, 100, 10). The number of viral particles bound per cell was calculated using the virion specific radioactivity and the number of cells.

### 3.2.7. Adenovirus-/antibody-labeling with Cy3

Adenoviruses or the hexon antibody were labeled with Cy3 using the Cy3 bis-Reactive Dye (Amersham). To constitute the labeling reagent, the dye was diluted in 0.1M sodium carbonate buffer (pH 9.3, Sigma). The ratio between the volumes of adenovirus or hexon antibody and labeling reagent was 1/9. The conjugation of the dye was carried out for 1 hour at room temperature in the dark, while the reaction tube was gently mixed (inverted) every 10 min. Labeled viruses were dialyzed against 10mM Tris-HCl, pH 7.5, 10mM $MgCl_2$, and 10% glycerol solution at 4°C overnight using a 10 kDa cut-off dialysis membrane cassette (Thermo Scientific) to remove unincorporated chemicals. The hexon antibody was dialyzed against PBS (pH 7.2, Sigma) at 4°C overnight using a similar cassette. Cy3-labeled virus was stored at -20°C, the Cy3-hexon antibody at 4°C.

### 3.2.8. Inhibitors

$5x10^4$ cells were seeded in 96well plates or 24well plates (BD Falcon) with tissue culture inserts (see 3.2.3). Confluent R/E and S/M cells were treated with inhibitors (Table 3.4) diluted in fresh MEGM medium overnight and then infected with Ad5/35.IR.E1A/TRAIL (100 pfu/cell). Four days after infection cells were subjected to an MTT assay and their metabolic activity was compared to appropriate controls (0.1% DMSO for Wortmannin/H-1152 or $dH_2O$).

### 3.2.9. Basal/apical infections with adenovirus

Tumor cells were infected with adenoviruses from the apical side using a total volume of 100µl on 96well, 200µl on 48well, 500µl on 24well or 2ml on 6well plates. On glass slides 500µl were used. For basal infection PET track-etched membrane tissue culture inserts with 1.0 µm pore size (BD Falcon) for 24well or 6 well plates were used with a total volume of 500µl (basal)/250µl (apical) or 2ml (basal)/1ml (apical), respectively. For quantitative readouts of GFP transduction (after Ad5/35.GFP or Ad5.GFP infection), samples were analyzed by flow cytometry at indicated time points (FACScan, Becton Dickinson).

### 3.2.10. MTT assay

An MTT stock solution of 5mg/ml 3-(4,5-Dimethylthiazol-2-yl)-2,5-diphenyltetrazolium bromide (Sigma) was constituted in PBS (Gibco) and stored at -20°C. MTT assays were carried out in 96 well plates (Fisher). At 4 or 8 days post infection 20µl of MTT stock solution (in PBS) was added to each well and cells were incubated for 2h at 37°C. Medium was removed, cells washed twice with PBS and air-dried. Then 100µl of DMSO (Sigma) was added per well and incubated for 30min at RT in order to dissolve crystals. Absorbance was measured in plate reader (EL 340, Bio-Tek Instruments) at 546nm. The unspecific background was subtracted from samples using Excel 2007 (Microsoft).

### 3.2.11. Cytopathic effect assay

For crystal violet staining, cells were fixed 8 days after infection with 4% paraformaldehyde (Sigma) for 10 min at room temperature. Fixed cells were incubated for 10 min in 1% crystal

## Material and Methods

violet (Sigma) in 70% ethanol, followed by three rinses with water. Photos were taken with a Lumix camera (Panasonic).

### 3.2.12. DNA Expression Arrays

Clonal ovc316m cultures were seeded in 96well plates (6 wells/clonal culture, BD Falcon) and 1 well of a 24well plate (BD Falcon). Two days later 3 wells in 96well plates were infected with Ad5/35.IR-E1A/TRAIL (MOI 100 pfu/cell) at 100% confluence, the other 3 wells were control infected (using PBS). 8 days post infection MTT assay was carried out as described above (3.2.8.) to determine resistant and susceptible cultures. Uninfected tumor cells in 24well plates were lysed in RLT buffer (Qiagen) and stored at 80°C until further use. Total RNA of 15 resistant and 16 susceptible cultures was prepared using RNAeasy columns (Qiagen) its quality assessed by Agilent 2100 bioanalyzer. RNA extracted from SKOV3-ip1 cells was used as a reference. mRNA fractions were amplified by linear *in vitro* transcription using the Amino Allyl MessageAmp aRNA kit (Ambion) and the resulting aRNA was coupled with Cy5 and Cy3 (31 samples, 62 chips using dye swap) and hybridized to Human Oligonucleotide 22K spotted microarrays at the Fred Hutchinson Cancer Research Center microarray facility, scanned and data were extracted using GenePix Pro 3.0 software (Molecular Devices). Microarray data were normalized using the LOWESS algorithm and processed in Bioconductor Limma package (Smyth, 2004). This package includes pre-processing capabilities for two-color spotted arrays. The differential expression methods apply to all array platforms and treat Affymetrix, single channel and two channel experiments in a unified way. The used R-code was provided by David Pritchard (University of Washington). The mean fold-change of data from dye swaps was generated with Microsoft Excel 2007 (Microsoft). Genes with less than 34% of recognized intensities (above background) throughout samples and below a fold-change of 1.2 between resistant and susceptible groups were excluded from further processing. TIGR Multiexperiment Viewer software (Saeed et al., 2003) was utilized for hierarchical clustering of resistant and susceptible cultures with these criteria. The resulting list of 983 differentially expressed genes was further processed in Pathway-Express (Khatri et al., 2006). Here, the gene symbol and corresponding fold-changes of genes were used with pre-set options. Original cpr-files and normalized data are stored at the NCBI Gene Expression Omnibus under accession numbers GPL8315 (platform) and GSE15473 (superseries).

### 3.2.13. qPCR for viral genomes

Confluent cells of three different R/E or S/M cultures were infected with Ad5/35.IR.E1A/TRAIL in an MOI of 100 pfu/cell in triplicates. The experiment was stopped 3 hours or 3 days post infection. Cells were washed with PBS, trypsinized, washed with RPMI+10% FBS and washed again with PBS twice. Nucleic acids from cell pellets were isolated using the AllPrep DNA/RNA Mini Kit (Qiagen) according to the manufactorer's protocol. 1ng DNA per reaction was used for further analysis. A standard curve for genomic DNA (isolated from SKOV3-ip1 cells) was prepared based on the equation that one copy number equals 3pg of genomic human DNA (6pg/diploid cell). Primers against Ad5 hexon were used to assess the viral copy number in viral genome preparations from fixed numbers of viral particles (assessed spectrophotometrically), serially diluted and spiked with 1ng genomic DNA of SKOV3-ip1 cells to generate a standard curve. All reactions were performed in triplicate in a total reaction volume of 15μl using ImmoMix (Bioline), SYBR green (Bioline) and 3 μmol/l of each primer, and carried out in the GeneAmp 7900 instrument (Applied Biosystems). The following parameters were used for amplification and melting curve analysis: 95°C (10 minutes), followed by 40 cycles of 60°C (2 minutes), 95°C (15 seconds), 60°C (15 seconds), 95°C (15 seconds). Ct values were calculated using the Sequence Detection System software (Applied Biosystems). Under these conditions at least 10 copies could be detected for each replicate. Levels of hexon were standardized per DNA genomic copy number using primers against housekeeping gene HPRT1. All primer sequences are supplied in Tables 3.6 and 3.8.

### 3.2.14. qRT-PCR for viral gene expression

Total RNA from the same experiment as described above was quality-analyzed by Agilent 2100 bioanalyzer, DNase treated (Invitrogen) and then transcribed into cDNA using Superscript III (Invitrogen). Therefore, 1μg of total RNA was mixed with 5μM oligo(dT) primers (Invitrogen) and 1mM dNTP mix (Invitrogen) in a total volume of 10μl. Samples were incubated for 5 min at 65°C and then placed on ice. For cDNA synthesis the following 10μl mix was applied to samples:

        2μl  10x RT buffer
        4μl  25mM $MgCl_2$
        2μl  0.1M Dithiothreitol (DTT)
        1μl  RNaseOUT (40 U/μl)
        1μl  SuperScript III reverse transcriptase (200 U/μl)

The following reactions were carried out in a PCR machine (Eppendorf, Mastercycler Gradient). Samples were incubated at 50°C for 50 min, 85°C for 5 min, and left at 4°C for at least 5 min. Tubes were centrifuged (1 min, 14,000xg) and then 1μl RNase H (Invitrogen) was added. Samples were then incubated at 37°C for 20 min and stored at -20°C until further use. qPCR procedures including reaction preparation, PCR procedure and CT value calculation were carried out as described above (3.2.11). Copy numbers of viral genes were standardized to HPRT1 expression.

### 3.2.15. qRT-PCR for array validation

Total RNA from R/E (N=5) and S/M (N=5) clones used for DNA expression analysis was utilized to synthesize cDNA and subsequent procedures were carried out as described above (3.2.13). The qPCR fold-changes shown in Table 4.1 represent a mean of data that were standardized against housekeeping genes Cyclophilin A and GAPDH (see Tables 3.7 and 3.8 for primer sequences). Individual fold-changes for clonal cultures are supplied in Supplementary Figure 7.5.

### 3.2.16. Immunofluorescence/Confocal microscopy

Cells were cultured in 8 chamber glass slides (BD Falcon), washed twice with ice-cold PBS and then fixed with methanol/aceton (1:1 vol/vol) for 15 min at 4°C. For confocal microscopy and actin staining (Phalloidin) cells were fixed with 4% paraformaldehyde for 30 min at 4°C. After fixation cells were washed with PBS twice and blocked with 500 µl PBS/2% dry-milk powder for 20 min at room temperature. Antibody staining was performed in 100 µl PBS for 90 min at 37°C or 4°C overnight. All antibodies and the used concentrations are provided in Table 3.1. If needed, fluorophor-labeled secondary antibodies directed against the appropriate host, were applied after 3 washes with PBS for 45 min at room temperature. After 3 washes with PBS glass slides were mounted using VECTASHIELD with DAPI (Vector Labs) or PBS/Glycerol (1:1 vol/vol, Gibco/Sigma) for confocal microscopy. DAPI staining (1:1000) was additionally included in the secondary antibody incubation, when needed. All immunofluorescence pictures were taken with a Leica DM1000 microscope featuring a Leica DFC FX camera (Leica Microsystems). Confocal pictures were generated using the Zeiss LSM 510 META system. The Argon ion and Diode lasers were used with the plan-apochromat 63x/1.4 DIC oil immersion objective. The x-y-z settings for individual z-scan pictures varied between 51µm x 0.1µm x 8.7µm to 70.8µm x 0.1µm x 12.3µm.

### 3.2.17. Immunohistochemistry on tumor sections

Tumor sections of patient material or xenografts were snap frozen embedded in OCT compound (Tissue-Tek) on dry ice. OCT embedded tissues were then stored at -80°C and equilibrated to -20°C for at least 1 hour prior to sectioning. Tumor tissue was sliced (8 microns) using the Leica CM 1850 cryostat (Leica Microsystems) and then transferred onto superfrost slides (Fisher Scientific). Slides were fixed in acetone (Fisher Scientific) for 10 min at -20°C. After two rinses with PBS (Sigma) slides were blocked with 2% milk powder (BioRad) in PBS for 20 min at room temperature. Further procedure was carried out as described for cultured cell immunofluorescence staining above (3.2.14).

### 3.2.18. X-gal staining on tumor sections

In order to assess β–galactosidase expression after intravenous injection of Ad5/35.β–gal, tumor sections were prepared using the cryostat (Leica Microsystems) as mentioned in 3.2.15. Sections were then fixed with ice-cold 0.5% glutaraldehyde in PBS (Sigma) for 15 min at 4°C. After two washes with PBS, X-gal staining solution (listed below) was added and slides were incubated at 37°C for 12-16 hours.

X-gal staining solution: 1.3 mM $MgCl_2$ (Sigma), 15 mM NaCl (Sigma), 50 mM HEPES pH 7.4 (Sigma), 3 mM potassium ferricyanide (Sigma), 3 mM potassium ferrocyanide (Sigma), 0.1% X-gal solution (5-bromo-4-chloro-3-indolyl-β-D-galactoside in dimethylformamide) (Sigma/VWR)

### 3.2.19. Flow cytometry

Cells were isolated from tumor xenografts as described above (3.2.1.). Adherent cells were detached from tissue culture plates by treatment with versene (Gibco) for 30-60 min. Cells were then transferred into 15ml blue cap tubes (BD Falcon) and washed with RPMI1640 supplemented with 10% FBS. Cell pellets were resuspended in ice-cold PBS with 1% FBS in order to block unspecific antibody binding. $2 \times 10^5$ cells were incubated with antibodies in 5ml round bottom tubes (BD Falcon) in a total of 100µl for 45 min on ice. All subsequent incubation steps were carried out in the dark. Cells were washed with 3 ml PBS+1%FBS and centrifuged at 400xg for 5 min at 4°C in between. After surface antigen staining, cells were fixed with 4% paraformaldehyde for 15 min on ice. Following a PBS+1%FBS wash, cells were either subjected to flow cytometry analysis or prepared for intracellular antigen staining (vimentin, N-p120) by treatment with 0.1% TritonX100 (Sigma) for 15 min at room temperature. For N-p120 staining, a PE-labeled anti-rabbit antibody was applied after 3 washes with 3ml PBS+1% FBS for 30 min at room temperature. Samples were filtered (0.22 µm, Becton Dickinson) and then analyzed using the BD FACSCanto™ flow cytometer (Becton Dickenson). Unspecific background of individual channels was determined with isotype controls and color compensation was done on single color-stained samples. Single cell gating was achieved by plotting forward-scatter-height versus forward-scatter-width. Figures were generated using FlowJo 8.7 for Macintosh (Tree Star, Inc.).

### 3.2.20. Calcium depletion by versene/ZOT treatment

Resistant clonal cultures were cultured in 96well plates or 8well chamber glass slides (BD Falcon). For calcium depletion cells were washed with PBS (Gibco) and then incubated with 100µl (96well plate) or 250µl (glass slide) versene (Gibco) for 3 hours at 37°C. Then, versene was removed and transferred into 1.5ml eppendorf tubes containing 1ml RMPI with 10% FBS (Gibco) for pelleting of detached cells (5 min, 400xg, Eppendorf table centrifuge). Immediately after versene-removal, 75µl (96well plate) or 475µl (glass slide) MEGM medium was applied to cells and the pelleted, detached cells were then re-applied using the same medium. Cultures were infected with 25 µl Ad5/35.IR-E1A/TRAIL dilution or PBS (as a control). For ZOT treatment, 4µg/ml ZOT were applied for two hours prior to infections with Adenoviruses. Control cells were treated with ZOT elution buffer (see 3.1.7.). ZOT-/buffer-containing medium was removed and cells were infected with indicated amounts of adenoviruses.

### 3.2.21. Animal studies

To test tumoriginicity of primary ovarian cancer cells, $1\text{-}2 \times 10^6$ cells were injected with matrigel (1:1 vol/vol, 120µl total volume) into the mammary fat pad of CB17-SCID-beige mice. Human ovarian cancer cells derived from xenografts were isolated as described above and then cultured in MEGM supplemented with 10% FBS. To establish subcutaneous tumors from ovc316 cells and other indicated cell lines, if not otherwise mentioned, CB17-SCID-beige mice were injected into the mammary fat pad with $1 \times 10^5$ tumor cells in matrigel (1:1 vol/vol, 120µl total volume). $2 \times 10^9$ pfu of adenoviral vectors were injected intratumorally when tumors reached a diameter of 5 mm. For intravenous application, $2 \times 10^9$ pfu of adenovirus were injected in 100µl of PBS through the tail vein.

### 3.2.22. Rhotekin/PAK GTPase pull-down

Cells were cultured in 6 well plates using tissue culture inserts. Tissue culture inserts were placed in 6 well plates containing either virus in MEBM or fresh MEGM supplemented with 10% FBS. After 15 min cells were placed on ice, washed with cold PBS twice and then lysed in protein lysis buffer. GTP-carrying forms of Rho GTPases were isolated from 500µg protein of fresh lysates using the Rho/Rac/Cdc42 Activation Assay Kit from Cell Biolabs. Two aliquots of individual samples were brought up to a total volume of 1ml with 1X Assay Lysis Buffer in 1.5ml eppendorf tubes. 40 µL of either resuspended Rhotekin RBD (for Rho pull-down) or PAK PBD (for Rac/Cdc42 pull-down) Agarose bead slurry was added to idividual aliquots. The samples were incubated for 1 hour at 4°C in a rotating shaker and beads then pelleted by centrifugation for 10 seconds at 14,000xg (Eppendorf table centrifuge). Following three wash/centrifugation steps with 0.5 ml of 1X Assay Buffer the pellet was resuspended in 40 µL of 2X reducing SDS-PAGE sample buffer (4% SDS, 20% glycerol, 10% 2-mercaptoethanol, 0.004% bromphenol blue and 0.125 M Tris HCl, pH 6.8 [Sigma]). 20µl of resulting samples were used for Western Blot analysis.

### 3.2.23. Western Blot

Xenograft tumor tissue was dissected, manually homogenized (glass homogenizer) and incubated for 30 min in protein lysis buffer (20mM Hepes (pH 7.5), 2mM EGTA, 10% glycerol, 1% TritonX100, 1mM PMSF, 200µM $Na_3VO_4$ [all Sigma] and protease inhibitors [Complete Protease Inhibitor Cocktail, Roche]) on ice. Confluent cultured cells were washed with ice-cold PBS twice and then lysed for 30 min in protein lysis buffer on ice. After 30 seconds sonication (Branson Sonifier 250, intensity 4 of 10) on ice, samples were pelleted (10 min, 4°C, 15,000 RPM, Eppendorf table centrifuge) and protein containing supernatant stored at -80°C. 15µg of total protein was used for western blotting. Protein samples were boiled (5 min, 95°C) and separated by polyacrylamide gel electrophoresis using 4-15% gradient gels (BioRad) followed by transfer onto nitrocellulose membranes according to the supplier's protocol (Mini ProteanIII, BioRad). Membranes were blocked in PBS, 0.1% Tween20 (PBS-T, Sigma) +5% dry milk powder (BioRad). Incubation times for primary and secondary antibodies were 2 hours and 1 hour at room temperature, respectively. Antibodies were diluted in PBS-T +2% dry-milk powder. Membranes were washed 5 times in PBS-T between antibody incubations and films were developed using ECL plus (Amersham). Antibodies and used concentrations are listed in Table 3.3.

### 3.2.24. Statistical analysis

All results in bar charts are expressed as mean of indicated replicates +/- standard deviation. Statistical significance was evaluated using Prism version 4.00c for Macintosh (GraphPad Software). Student's T-test or ANOVA (analysis of variance) for multiple testing were applied when applicable. A p value <0.05 was considered significant.

## 3.3. Suppliers

| Company | Location |
|---|---|
| Abcam | Cambridge, UK |
| American Type Culture Collection (ATCC) | Manassas, VA, USA |
| Amersham | Little Chalfont, UK |
| Applied Biosystems | Foster City, CA, USA |
| Baxter | McGaw Park, IL, USA |
| BD Falcon | San Jose, CA, USA |
| BD Pharmingen | Mississauga, ON, Canada |
| BD Transduction Laboratories | Franklin Lakes, NJ, USA |
| Beckman Coulter | Fullerton, CA, USA |
| Becton Dickinson | Franklin Lakes, NJ, USA |
| Bio-Tek Instruments | Winooski, VT, USA |
| Bioline | Taunton, MA, USA |
| Biomeda | Foster City, CA, USA |
| BioRad | Hercules, CA, USA |
| Calbiochem | Nottingham, UK |
| Cell Biolabs | San Diego, CA, USA |
| Cell Signaling | Danvers, MA, USA |
| CellGro | Manassas, VA, USA |
| Charles River Laboratories | Wilmington, MA, USA |
| Chemicon | Temecula, CA, USA |
| Dako | Glostrup, Denmark |
| Fisher Scientific | Pittsburgh, PA, USA |
| Fitzgerald | Concord, MA, USA |
| Genetex | Irvine, CA, USA |
| Gibco | Carlsbad, CA, USA |
| GraphPad Software | San Diego, CA, USA |
| Greiner | Monroe, NC, USA |
| Invitrogen | Carlsbad, CA, USA |
| InvivoGen | San Diego, CA, USA |
| Leica Microsystems | Wetzlar, Germany |
| Lonza | Basel, Switzerland |
| Microsoft | Redmond, WA, USA |
| Nalgene | Rochester, NY, USA |
| Operon | Huntsville, AL, USA |
| Panasonic | Secaucus, NJ, USA |
| Qiagen | Valencia, CA, USA |
| R&D Systems | Minneapolis, MN, USA |
| Roche | Mannheim, Germany |
| Sigma | St. Louis, MO, USA |
| Southern Biotech | Birmingham, AL, USA |
| Spectrum Laboratories | Rancho Dominguez, CA, USA |
| Stem Cell Technologies | Vancouver, Canada |
| Thermo Scientific | Rockford, IL, USA |
| Tissue-Tek | Torrance, CA, USA |
| Tree Star, Inc. | Ashland, OR, USA |
| Vector Labs | Burlingame, CA, USA |
| VWR | West Chester, PA, USA |
| Zymed | San Francisco, CA, USA |

# 4. Results

## 4.1. Screening of primary ovarian cancer cultures for resistance to viral oncolysis

As a result of the genetic instability of tumor cells in general and the conditions within the tumor microenvironment, cell populations in solid tumors are extremely heterogeneous. Considering this, it was hypothesized that primary tumor cultures are a more appropriate model to analyze the performance of oncolytic adenoviruses when compared to established cancer cell lines. Specifically, it was hoped that this cell diversity might give rise to sub-populations that can escape viral oncolysis. A total of 42 primary cultures were established from tumor biopsies or ascites of stage III and IV ovarian cancer patients. Cells in early passages (passage 4 or less) were then infected with an oncolytic adenovirus (Ad5/35.IR-E1A/TRAIL) at MOIs ranging from 1 to 100 plaque forming units (pfu)/cell. This vector is targeted to CD46 by the adenovirus serotype 35 fiber, allows for tumor-specific, replication-activated expression of E1A and TRAIL, and efficient tumor cell killing (see Fig. 1.8). In contrast to an otherwise similar oncolytic vector based on serotype 5 that uses CAR as primary receptor, infection with Ad5/35.IR-E1A/TRAIL vector caused cell death in all tumor cultures at day 4 after infection. Interestingly, in about 10% (N=4) of tested ovarian cancer cultures, individual cells or cell clusters remained alive, even when high doses of Ad5/35.IR-E1A/TRAIL virus were applied (Fig. 4.1). Notably, 100% cell lysis was observed for SKOV3-ip1, an established cell line that is widely used in ovarian cancer models.

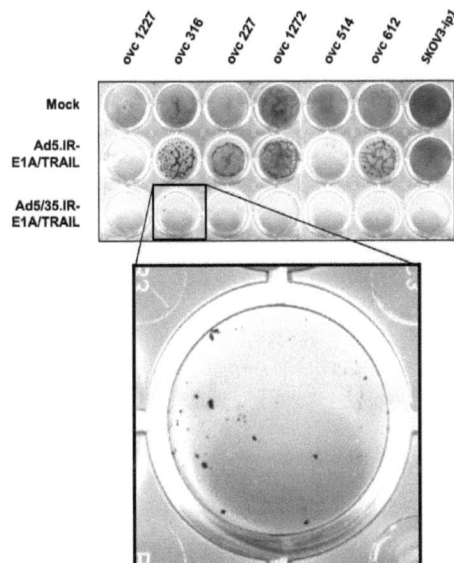

Figure 4.1: Cytolysis of primary ovarian cancer cultures. A total of $5 \times 10^4$ cells of early passage (less than p4) primary ovarian cancer cells as well as SKOV3-ip1 cells were seeded per well in 24 wells the next day infected with the Ad5 and Ad5/35-based oncolytic vectors (MOI 50 pfu/cell). Cells were stained with crystal violet 4 days later. Note that cell clusters survive treatment with Ad5/35.IR-E1A/TRAIL within culture ovc316. A representative series of cultures is shown.

Further studies concentrated on culture ovc316, which was obtained from a stage IVB serous ovarian cancer that also showed resistance to primary chemotherapy. In contrast to other tested primary cell lines, this culture was capable of forming tumors in CB17-SCID-beige mice when $10^6$ cells were injected. Additionally, further analysis of other cultures that contained cell subsets resistant to viral oncolysis was compromised by early cell senescence (passages ≤7). Cells isolated from an ovc316 xenograft tumor (termed ovc316m) were heterogeneous (Fig. 4.2B), free from fibroblasts, and showed unlimited proliferative potential. In order to enrich for cells resistant to viral oncolysis, 120 single cell clones were established by limited dilution culturing of ovc316m cells that were passaged three times (Fig. 4.2A). The morphology of uninfected clonal cultures from ovc316 varied greatly. About 40% of clones contained relatively homogenous cell populations (Fig. 4.2C). However, the majority of clones developed clearly distinguishable subsets of cells with different morphology within a given culture, indicating that the cell the clone derived from was pluripotent.

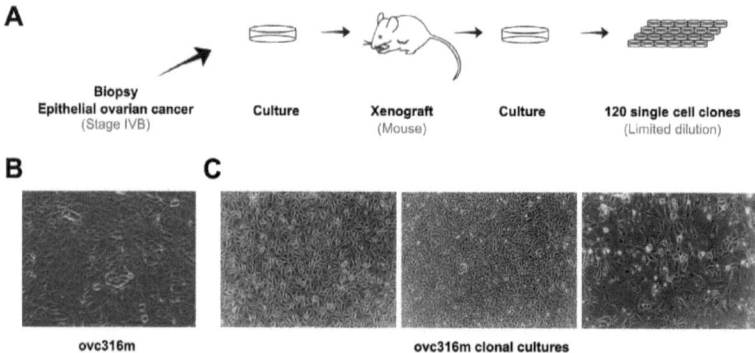

Figure 4.2: Model for the enrichment of cells that are resistant to viral oncolysis. A) Patient biopsies were adapted to tissue culture and tested for resistance against adenoviral oncolysis. Cultures containing resistant cell subsets were injected into immunocompromised mice and subjected to clonal cell culturing. B/C) Bright field microscopy of primary ovarian cancer culture ovc316. B) Cells isolated from mouse xenograft (ovc316m) show high heterogeneity with cell subsets of different morphology. C) Clonal cultures derived from culture ovc316m differ greatly in their morphology.

To identify resistant cultures, a total of 100 ovc316m clones in passage 6 were then infected with Ad5/35.IR-E1A/TRAIL at an MOI of 100 pfu/cell (at 100% cell confluence). Metabolic activity and cell viability was analyzed 8 days post-infection by MTT assay (Fig. 4.3 and 4.4A). Resistance was determined by intact cell monolayers. The metabolic activity of these cultures varied between 65-90%. 20% of clonal cultures appeared to be resistant, while 19% were susceptible to oncolysis (<10% metabolic ativity). The remaining cultures contained between 10 and 65% viable cells. Infection at MOIs 10 and 200 pfu/cell resulted in a similar distribution (data not shown). These data indicate that the morphological heterogeneity seen in clonal cultures is also reflected in a heterogenic response to the infection by Ad5/35.IR-E1A/TRAIL.

# RESULTS

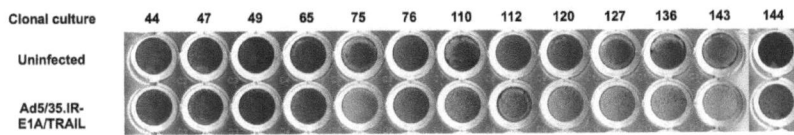

Figure 4.3: Screening of clonal ovc316m cultures for resistance to viral oncolysis. Metabolic activity of clonal ovc316m cultures 8 days after treatment with Ad5/35.IR-E1A/TRAIL (MOI 100 pfu/cell) measured by MTT assay. Shown is a representative series of clonal ovc316m cultures before formazan crystals were dissolved for MTT assay readout. Experiments were carried out in triplicates. Cultures with an intact monolayer were defined as resistant, cultures with less than 10% cell survival determined as susceptible to viral oncolysis.

## 4.2. Resistant ovarian cancer cells have an epithelial phenotype

Clonal cultures that were resistant to Ad5/35.IR-E1A/TRAIL lysis (N=15) were subjected to genome-wide mRNA expression analysis in comparison to clonal cultures susceptible to viral oncolysis (N=16). Using the Bioconductor (Linear Models for Microarray Data) software package (Smyth, 2004), 983 genes were found to be differently expressed (p<0.017). Hierarchical clustering of these genes showed a clear-cut separation of resistant and susceptible clones (Fig. 4.4B). This level of discrimination is usually not achievable when populations of primary cancer cultures are used. Among the top genes downregulated in resistant clones were *IL-6, vimentin, VCAM-1, CD44, FGFR1, palladin, caldesmon* and *vinculin*. Among the top upregulated genes in resistant clones were *epithelial cell adhesion molecule (EpCAM), claudins 3, 4* and *7, CDH1 (E-cadherin), netrin 4* and *mucin1*. Additionally, the gene ontology software tool "Pathway-Express" (Khatri et al., 2006) was utilized to identify potentially affected cellular pathways. This analysis revealed pathways involving tight and adherens junction formation and cell adhesion, as well as pathways involved in antigen presentation/processing, were significantly different in resistant cells (p<0.008) (Fig. 4.4C). The antigen presentation/processing involvement was disregarded, because of the absence of immune cells in the chosen experimental setup. Further studies focused on the tight junction and cell adherens pathways. Figure 4.4D shows key members of tight and adherens junction pathways and their expression status in resistant clones. Altered RNA expression levels of 26 selected genes found in microarray studies and other members of these pathways were validated by qRT-PCR (Table 4.1 and Suppl. Fig. 5). Gene expression on protein level was studied by immunofluorescence and flow cytometry analyses (Fig. 4.5). Immunofluorescence analyses demonstrated high-levels of adherens proteins (E-cadherin, β-catenin), tight junction proteins (occludin, claudin 3, 4, 7, cingulin), and the epithelial marker EpCAM in resistant cells (Fig. 4.5A). Conversely, susceptible cells predominantly expressed markers that are characteristic for mesenchymal cells (vimentin, vinculin, N-cadherin) (Fig. 4.5B). Furthermore, expression of CD44 was slightly higher in susceptible than resistant clones. Notably, CD44 is considered a marker for mesenchymal stem cells (Barry and Murphy, 2004; Conget and Minguell, 1999; Pittenger et al., 1999). Flow cytometry analysis revealed that the majority of cells in resistant clones express high levels of EpCAM, but also the mesenchymal marker vimentin and CD44 (Fig. 4.5C). On susceptible cells, EpCAM levels are about 10-fold lower while the expression of mesenchymal markers is higher than on resistant cells (Fig. 4.5D).

# RESULTS

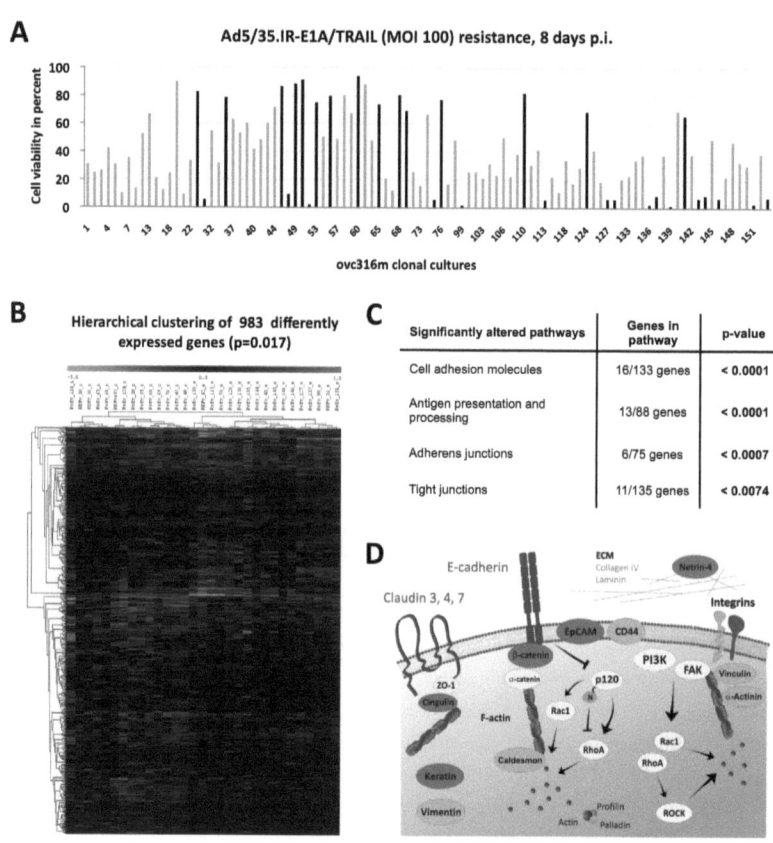

**Figure 4.4: Identification of altered pathways by DNA expression array analysis of clonal ovc316m cultures resistant or susceptible to viral oncolysis. A)** Metabolic activity of clonal ovc316m cultures 8 days after treatment with Ad5/35.IR-E1A/TRAIL (MOI 100 pfu/cell) measured by MTT assay. Cultures with an intact monolayer were defined as resistant, cultures with less than 10% cell survival determined as susceptible to viral oncolysis. Standard deviation of triplicates was less than 10%. Cultures indicated in black were chosen for subsequent DNA expression analysis. **B)** Hierarchical clustering of 983 significantly altered (>1.5fold change, p=0.017) genes, comparing 15 resistant and 16 susceptible cultures. Interestingly, the resistant culture with the lowest metabolic activity (65%) groups with susceptible cultures. **C)** Gene ontology analysis of 983 differently expressed genes showing significantly altered pathways, including the number of altered genes and the statistical significance. **D)** Focal adhesion, tight junction, and adherens junction pathways. Red, genes that were found up-regulated in arrays; green, down-regulated genes. Tight junction proteins include claudins and occludin. Adherens junction proteins include E-cadherin. The cytoplasmic domain of E-cadherin interacts with β-catenin and p120-catenin. The effect of the RhoA interaction with p120-catenin depends on the presence of the regulatory N-terminal domain. Claudins and occludin interact with ZO-1 (zonula occludens-1) and subsequently with F-actin via cingulin. caldesmon inhibits Arp2/3-mediated actin polymerization. Phosphoinositide 3-kinase (PI3K) and focal adhesion kinase (FAK) are involved in the regulation of Rho-GTPases downstream of integrin signaling. Rho kinase (ROCK) is activated by RhoA. Vinculin and α-actin crosslink the cytoskeleton to focal adhesion spots. Vimentin is the major intermediate filament protein of mesenchymal cells that is involved in regulation of attachment, migration, and cell signaling. Palladin functions as a scaffold that regulates actin organization. Profilin is involved in turnover of the actin filament network. Extracellular matrix (ECM) proteins are connected to cells via integrins and include laminin and collagen IV. Netrin-4 can interrupt Laminin networks.

| Gene Symbol | UniGene ID | Description | Fold change DNA array | Fold change qRT-PCR | p-value |
|---|---|---|---|---|---|
| ACTN1 | 119000 | Actinin, alpha 1 | 0.66 | 0.77 | 0.1881 |
| ARHA (RhoA) | 77273 | Ras homolog gene family, member A | 0.79 | 1.07 | 0.9064 |
| CDC42 | 146409 | Cell division cycle 42 (GTP binding protein, 25kD) | 0.95 | 2.51 | 0.1938 |
| CDH1 | 194657 | Cadherin 1, type 1, E-cadherin (epithelial) | 2.91 | 17.95 | 0.0012 |
| CFL1 | 180370 | Cofilin 1 (non-muscle) | 0.95 | 1.16 | 0.3557 |
| CLDN1 | 7327 | Claudin 1 | 0.73 | 0.65 | 0.3143 |
| CLDN2 | 16098 | Claudin 2 | 0.97 | 0.67 | 0.5690 |
| CLDN23 | 183617 | Claudin 23 | 1.00 | 2.77 | 0.0114 |
| CLDN3 | 25640 | Claudin 3 | 2.41 | 16.09 | 0.0021 |
| CLDN4 | 5372 | Claudin 4 | 1.60 | 17.48 | 0.0007 |
| CLDN7 | 278562 | Claudin 7 | 5.94 | 49.45 | 0.0003 |
| CTNNB1 | 171271 | Catenin (cadherin-associated protein), beta 1 (88kD) | 1.06 | 1.63 | 0.1582 |
| DIAPH1 (DIA1) | 26584 | Diaphanous homolog 1 (Drosophila) | 1.22 | 1.89 | 0.0036 |
| MAPK8 (JNK) | 190913 | Mitogen-activated protein kinase 8 | 1.02 | 1.37 | 0.2802 |
| MIC2 (CD99) | 177543 | Antigen identified by monoclonal antibodies 12E7, F21 and O13 | 0.67 | 0.75 | 0.1661 |
| MYH10 | 16355 | Myosin, heavy chain 10, non-muscle | 1.00 | 0.86 | 0.4494 |
| MYLK | 211582 | Myosin, light polypeptide kinase | 0.54 | 0.99 | 0.9229 |
| NTN4 | 102541 | Netrin 4 | 3.08 | 8.05 | 0.0065 |
| OCLN | 171952 | Occludin | 1.28 | 4.01 | 0.0072 |
| RAC1 | 173737 | Ras-related C3 botulinum toxin substrate 1 (rho family, small GTP binding protein Rac1) | 0.98 | 1.09 | 0.4012 |
| RAP1B | 156764 | RAP1B, member of RAS oncogene family | 0.81 | 1.10 | 0.6716 |
| ROCK1 | 17820 | Rho-associated, coiled-coil containing protein kinase 1 | 1.11 | 1.21 | 0.3164 |
| ROCK2 | 58617 | Rho-associated, coiled-coil containing protein kinase 2 | 0.85 | 1.82 | 0.0829 |
| RRAS2 (TC21) | 206097 | Related RAS viral (r-ras) oncogene homolog 2 | 0.46 | 0.47 | 0.0206 |
| VCAM1 | 109225 | Vascular cell adhesion molecule 1 | 0.31 | 0.10 | 0.0002 |

**Table 4.1: RNA expression levels found in DNA expression arrays validated by quantitative reverse transcriptase-PCR (qRT-PCR).** Shown is the fold change in mRNA levels between resistant and susceptible cultures (N=10, 5 selected from each group), visualized for the status in resistant cells. The official gene symbol (by Human Genome Organization (HUGO) Gene Nomenclature Committee) and Unigene identification number (ID) is indicated. Gene names are listed alphabetically. Significantly up-regulated genes (based on qRT-PCR) are shown in red, down-regulated genes are marked green (p<0.05).

In summary, clonal ovarian cancer cultures express both epithelial and mesenchymal markers, whereby clones resistant or susceptible to Ad5/35.IR-E1A/TRAIL show the balance greatly shifted towards epithelial or mesenchymal markers, respectively. In all studies described below, resistant and susceptible clones were designated as R/E (resistant/epithelial) and S/M (susceptible/mesenchymal), respectively. S/M cells proliferated about twice as fast as R/E cells. Furthermore, it was observed that R/E and S/M cultures also differed in their proliferative potential. Whereas S/M cultures became senescent between passages 20-25, R/E cultures could be passaged 25-50 times. Several resistant clonal cultures occasionally underwent an EMT and in part turned into S/M cells during further passaging in the standard growth medium (contains FBS, bovine pituitary extract, EGF, hydrocortisone and insulin). These cultures were labeled as R/E-EMT cells. In contrast, R/E cultures were completely refractory to EMT during passaging under the same conditions. A mesenchymal-epithelial transition (MET), where S/M cells would turn into R/E cells, was not observed in this study.

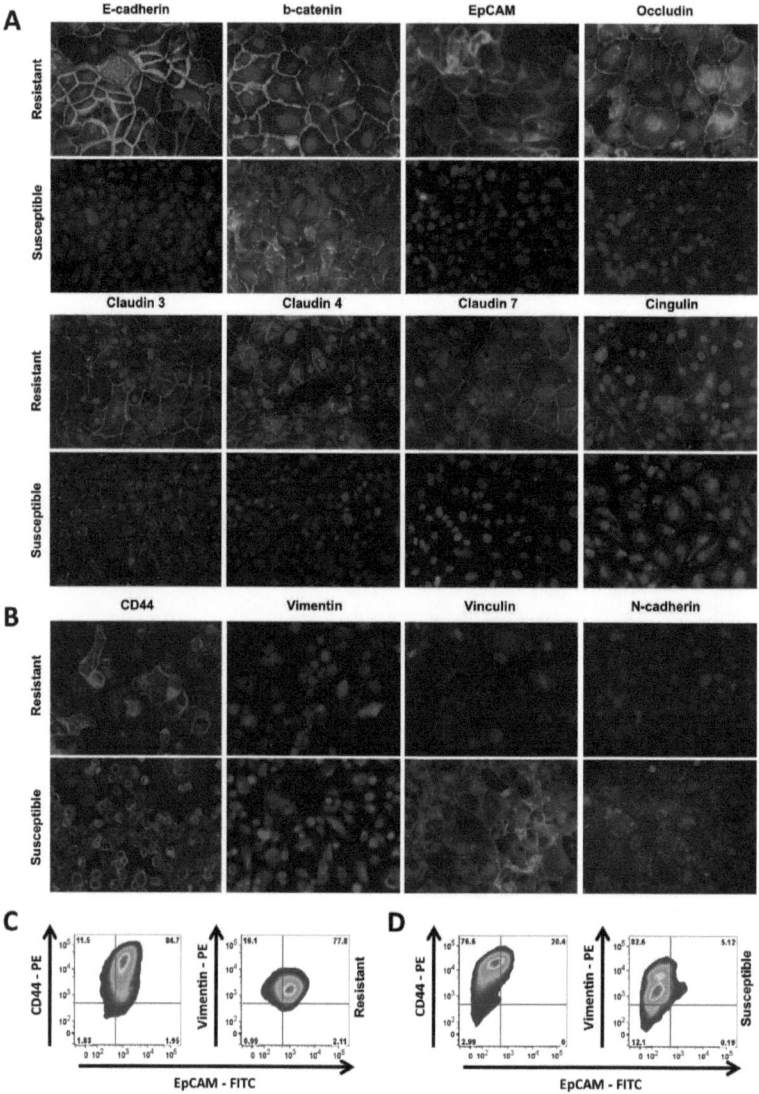

**Figure 4.5: Clonal ovc316m cultures resistant to adenoviral oncolysis have an epithelial phenotype. A/B)** Immunofluorescence analysis of ovc316m clonal cultures resistant or susceptible to viral oncolysis. Staining for epithelial marker proteins (A) and mesenchymal marker proteins (B). Resistant cells correlate with the expression of epithelial markers; susceptible cells have higher amounts of mesenchymal markers. Representative clonal cultures are shown. Nuclei are stained in blue. **C/D)** Flow cytometry analysis of resistant (C) and susceptible (D) clones. Two representative clones are shown. Resistant cells express epithelial marker EpCAM and mesenchymal marker vimentin as well as high amounts of mesenchymal stem cell marker CD44. Susceptible cells express higher amounts of mesenchymal marker vimentin and are almost completely negative for epithelial marker EpCAM.

The extracellular matrix components laminin, collagen IV, and fibronectin were predominantly expressed by S/M clones and detected as excessive protein networks on top of cell monolayers. In contrast, resistant cells stained positive for Netrin-4 (Fig. 4.6A), a protein that is related to Laminin. Figure 4.6B shows the actin cytoskeleton and members of intercellular junction complexes that did not significantly differ in their mRNA levels between R/E and S/M cultures. However, the distribution of the proteins appeared to be different. In S/M cells, F-actin and tight junction proteins claudin 1 and 2 were spread throughout the cytoplasm, whereas R/E cells primarily translocated these proteins to the lateral membranes.

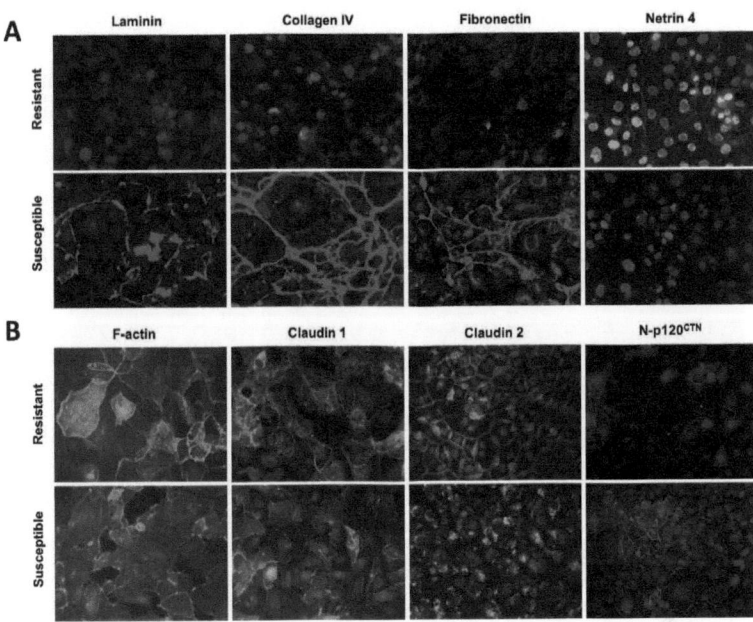

**Figure 4.6: Susceptible cells express extracellular matrix components and do not maintain tight junctions. A)** Excessive amounts of extracellular matrix proteins are located on top of susceptible cells. **B)** Proteins that are involved in identified cell adherens pathways, but not significantly different in mRNA expression levels. Note that N-p120 catenin is predominantly expressed in susceptible cells. Representative clonal cultures are shown. Nuclei are blue.

Of particular interest was the high prominence of p120 catenin containing the regulatory N-terminus in susceptible cells. To discriminate between different splice forms, an antibody targeting this N-terminus was used in this study. The detected "mesenchymal" transcription variant/s will be determined as N-p120 hereafter. The preferential expression of N-p120 is often accompanied with E-cadherin loss. In agreement with this, E-cadherin was expressed on the cell surface of R/E cells. In S/M cells this marker was either undetectable by immunofluorescence analysis or localized in the cytoplasm/nucleus (Fig. 4.7A). R/E-EMT cultures showed high levels of membrane bound N-p120 in cells negative for membrane-bound E-cadherin, but still retained large amounts of E-cadherin/claudin7 positive epithelial cells (Fig. 4.7B).

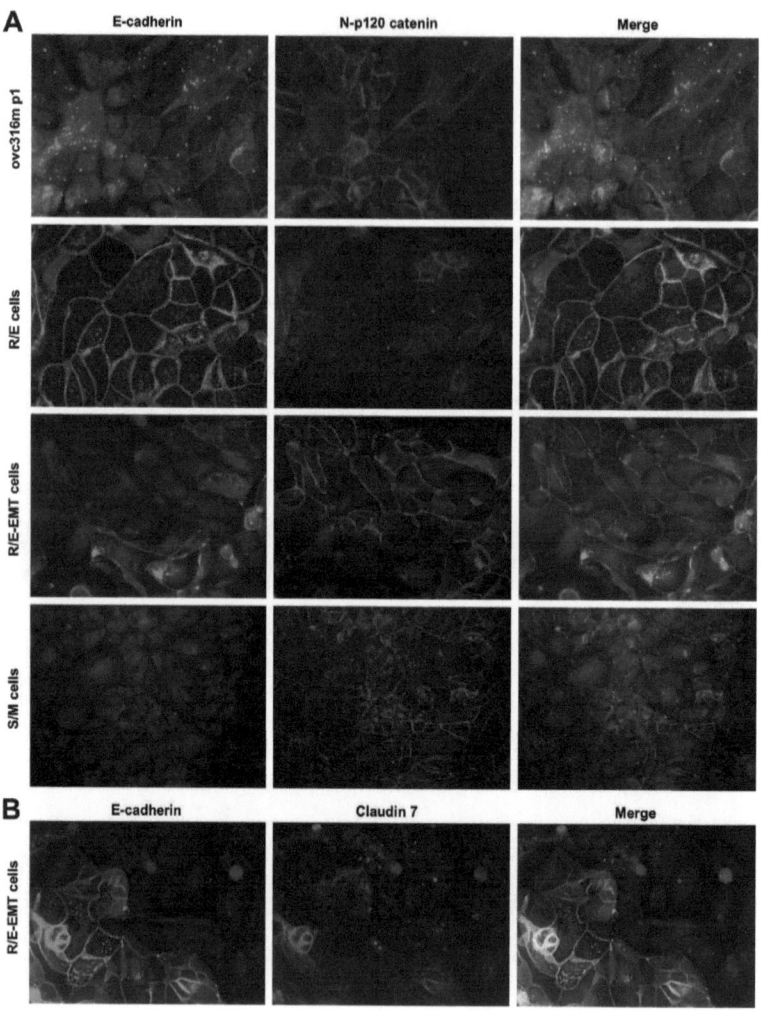

**Figure 4.7: Phenotypes of different ovc316-derived cultures identified in this study.** Immunofluorescence analysis of ovc316m cultures for E-cadherin and the regulatory N-terminal domain of p120 catenin **(A)**. R/E and S/M clonal cultures are derived from ovc316m cells (shown in passage 1). R/E-EMT cells are derived from R/E cells that underwent EMT during further passaging. Cells with E-cadherin staining on lateral membranes have low abundance of N-p120 and vice versa. In R/E cells, more intense staining for N-p120 correlates with stronger membrane staining for E-cadherin. Ovc316m p1 cells show an unorganized, punctated staining for E-cadherin. S/M cells have either lost E-cadherin or its localization is restricted to the cytoplasm/nucleus. R/E-EMT cells partly lost intercellular E-cadherin and gained N-p120, while retaining large amounts of cells that are E-cadherin/Claudin 7 positive **(B)**. Representative clonal cultures are shown. Nuclei are stained in blue.

Considering the genetic instability of cancer cells, the establishment of clonal cultures might have selected for certain cell features (e.g. proliferative capacity, independence of cell-cell contact, etc.). Therefore, it was tested whether the findings could be validated in the primary ovc316m culture from which the clones were derived. Notably, as Fig. 4.8 shows, the vast majority of cells in ovc316m cultures were susceptible to Ad5/35.IR-E1A/TRAIL. The amount of cells resistant to viral oncolysis varied between 10-20% in individual experiments at day 8 post infection (MTT assay).

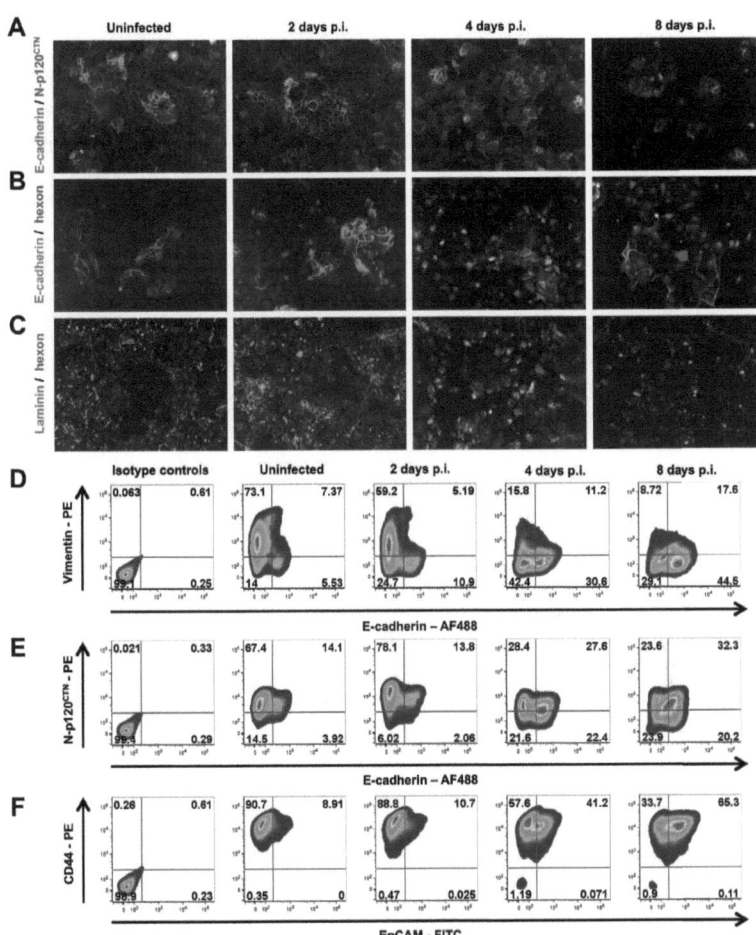

Figure 4.8: Epithelial cells within culture ovc316m survive treatment with Ad5/35.IR-E1A/TRAIL. A-F) Primary cultures (passage 10) were infected with Ad5/35.IR-E1A/TRAIL at an MOI of 100 pfu/cell and analyzed at days 2, 4, and 8 after infection by immunofluorescence (A-C) or flow cytometry (D-F). Uninfected cells were used as a control. Staining for E-cadherin and N-p120 (A), E-cadherin and viral hexon (B), and Laminin and viral hexon (C). Nuclei are stained in blue. D-F) Flow cytometry analyses show enrichment of epithelial populations (E-cadherin+/EpCAM+) in response to Ad5/35.IR-E1A/TRAIL. Representative samples are shown at 2, 4, and 8 days after infection.

Low-passage ovc316m cells were infected with Ad5/35.IR-E1A/TRAIL and analyzed for for *i)* the epithelial marker E-cadherin and the mesenchymal marker N-p120 (Fig. 4.8A), *ii)* viral hexon expression (a marker for viral replication) and E-cadherin (Fig. 4.8B), and *iii)* hexon and laminin (Fig. 4.8C). E-cadherin-positive and laminin-negative cells appeared to be resistant to lysis by Ad5/35.IR-E1A/TRAIL. At day 4 after infection most susceptible cells showed bright red viral hexon signals and started dying. At day 8 post infection, only E-cadherin positive cells remained in infected ovc316 cultures. Notably, resistant cells in clusters showed membrane localized E-cadherin, whereas individual cells that were not in contact to other cells demonstrated cytoplasmic E-cadherin signals. Furthermore, flow cytometry was utilized to monitor for E-cadherin/vimentin, E-cadherin/N-p120, and EpCAM/CD44 on infected cells over time (day 2, 4, 8 post infection) (Fig. 4.8D-F). While the percentage of vimentin$^{high}$ (mesenchymal) cells decreased from 80.5% (uninfected cells) to 64.4%, 27.0%, and 26.3% at days 2, 4, and 8 post infection, respectively, the percentage of E-cadherin$^{high}$ cells increased from 12.9% (uninfected), to 16.1% (day 2 post infection). 41.8% (day 4 post infection), and 52.1% (day 8 post infection). E-cadherin$^{high}$/vimentin$^{low}$ cells appeared to be more resistant than E-cadherin$^{high}$/vimentin$^{high}$ cells. Greater resistance of cells that differentiated towards epithelial cells is also seen in flow analysis of E-cadherin/N-p120 and EpCAM/CD44 (Figs. 4.8E/F). Notably, resistant cells that showed cytoplasmic E-cadherin staining in immunofluorescence studies scored negative by flow cytometry analysis for surface E-cadherin. In summary, cells that express epithelial markers are present at low frequency in primary ovarian cancer cultures and these cells are resistant to lysis by Ad5/35.IR-E1A/TRAIL.

## 4.3. Adenovirus receptors are trapped within tight junctions

Studies on ovc316 cells corroborated the findings obtained on clonal cultures. Further analysis to study the mechanisms of resistance to killing by Ad5/35.IR-E1A/TRAIL was therefore carried out on resistant and susceptible clones. To efficiently kill ovarian cancer cells, the oncolytic virus has to attach to cells, enter them, initiate viral genome replication, express TRAIL and E1A, and trigger apoptosis/cytolysis. First the attachment of $^3$H-labeled Ad5/35.IR-E1A/TRAIL particles was studied. Three times more adenovirus particles attached to S/M clones than to R/E clones (Fig. 4.9A). Consequently, subsequent infection steps are affected in R/E cells. The number of viral genomes found in cells at 3 hours post infection is about 10-fold lower in R/E cells (Fig. 4.9B). Within 72 hours after infection, due to viral genome replication, the difference between the amount of viral DNA in S/M and R/E cells was greater than 10,000 fold (Fig. 4.9B, right panel). Levels of mRNA for early (E2A) and late (pIX and hexon) viral genes, as well as replication-activated transcription of E1A and TRAIL transgenes, are markedly decreased in resistant clones (Fig. 4.9C). Moreover, using fluorophore-labeled (Cy3) Ad5/35 particles, an inefficient attachment to R/E cells was observed in immunofluorescence studies (Fig. 4.9D). These studies also revealed that the post-attachment signaling which results in re-organization of the F-actin network or recruitment of the focal adhesion protein vinculin in S/M clones, is not activated in R/E clones.

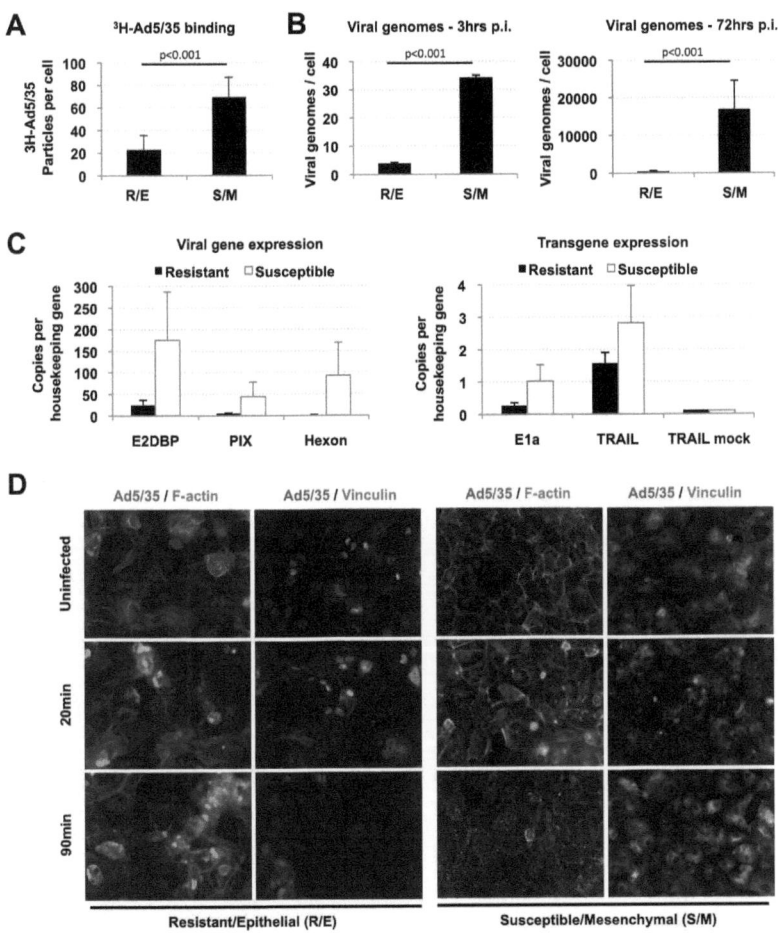

**Figure 4.9: Adenovirus infection steps are impaired in R/E cultures. A-C)** R/E and S/M cultures, N=10 (5 each). **A)** Attachment of 3H-labeled to cells. **B)** Virus uptake (left) and viral genome replication (right) analyzed by q-PCR. **C)** Expression of viral genes for E2-DNA-binding protein (E2DBP), protein IX (PIX), hexon (left) and transgene expression of E1a and TRAIL (right). **D)** Focal adhesion analysis in R/E and S/M clones. Cells were treated with 4000 Cy3-labeled Ad5/35 particles/cell and incubated for 20 min at 37°C. For the 90 min time-point cells were washed and incubated further. Cells were stained for F-actin cytoskeleton and focal adhesion marker vinculin. Nuclei are stained in blue.

Overall, these data suggest that inefficient attachment of adenovirus particles to epithelial ovarian cancer cells is largely responsible for resistance to killing by Ad5/35.IR-E1A/TRAIL. This implies that other vectors with an Ad5/35 capsid will also be inefficient in infecting epithelial ovarian cancer cells. The use of an Ad5/35 vector expressing GFP under the control of the CMV promoter (Ad5/35.GFP) could confirm this observation. The testing of all 31 R/E and S/M cultures, which were subjected to DNA expression array analysis, revealed a correlation in resistance to viral oncolysis and reduced GFP transduction in R/E cultures (Fig. 4.11A). Overall, transduction with Ad5/35-GFP was significantly less efficient in R/E clones than in S/M clones (Fig. 4.11B). GFP-positive cells in transduced E/R cell cultures were predominantly located on top of the monolayer (most likely mitotic cells) (Fig. 4.11C, upper panel), and at the rim of contact with the tissue culture plastic (Fig. 4.11C, central panel). Interestingly, lowered cell density did not proportionally increase Ad5/35-GFP transduction efficiency (Fig. 4.11D). Furthermore, a different oncolytic Ad5/35 vector (Ad5/35Δ24Ki/COX) was tested on ovc316m cells. This vector expresses the adenovirus E1 and E4 gene products under the control of the Ki67 and the cyclooxygenase-2 promoter, respectively [see also 1.4.2.3. (Hoffmann et al., 2008)]. Similar to Ad5/35.IR-E1A/TRAIL, Ad5/35Δ24Ki/COX predominantly killed mesenchymal cells (vimentin$^{high}$/p120$^{high}$) while epithelial cells (EpCAM$^{high}$, E-cadherin$^{high}$) were resistant (Fig. 4.10).

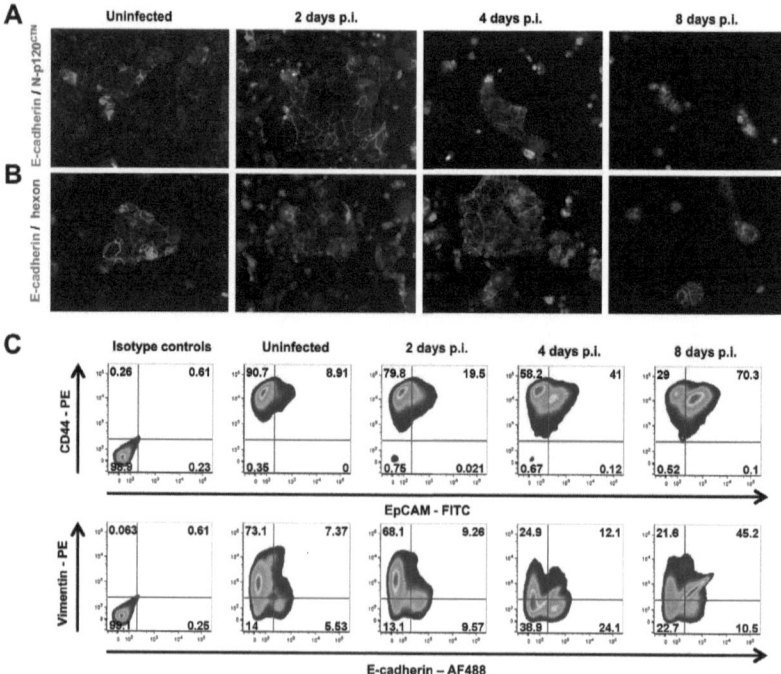

**Figure 4.10: Epithelial cells within culture ovc316m also survive treatment with oncolytic Ad5/35D24Ki/Cox.** A-C) Primary cultures (passage 10) were infected with Ad5/35Δ24Ki/COX at an MOI of 100 pfu/cell and analyzed at days 2, 4, and 8 after infection by immunofluorescence staining for E-cadherin and N-p120 **(A)**, and E-cadherin and viral hexon **(B)**. Nuclei are stained in blue. Uninfected cells were used as a control. **C)** Flow cytometry analysis of infected cells at days 2, 4, and 8 after infection. Representative samples are shown.

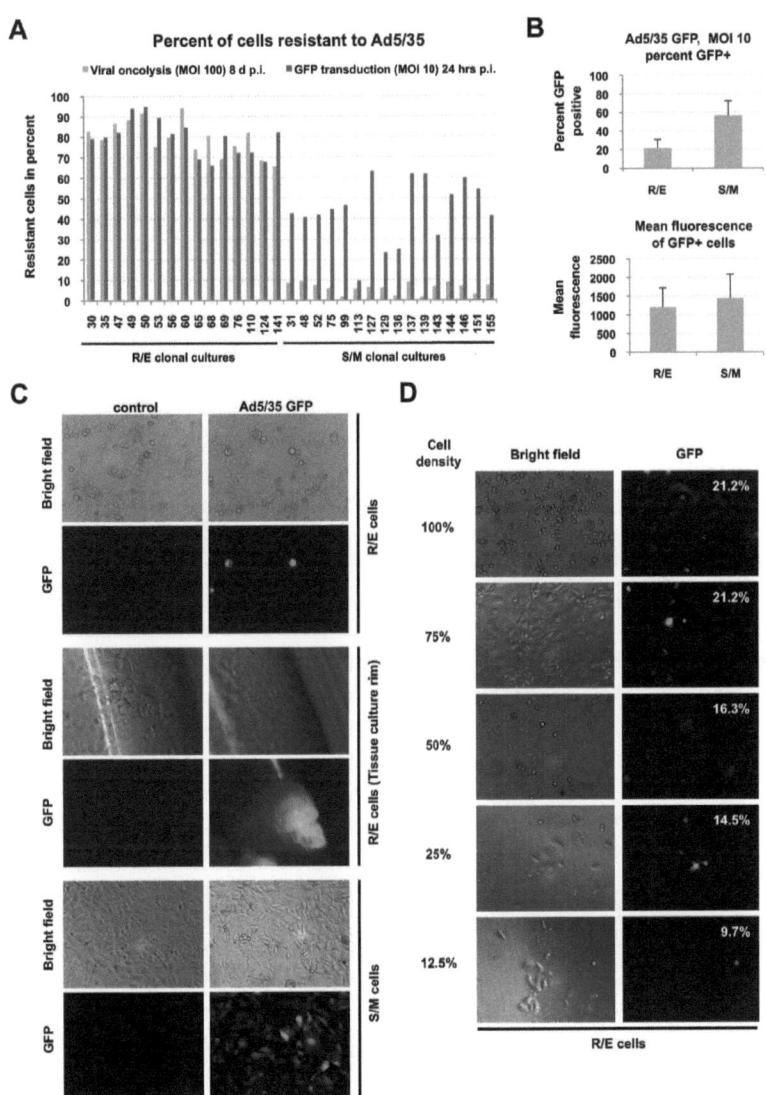

**Figure 4.11: Ad5/35 vectors achieve markedly less transduction in R/E cells. A)** Correlation of resistance to viral oncolysis (Ad5/35.IR-E1A/TRAIL) and resistance to GFP transduction (Ad5/35.GFP) in all clonal ovc316m cultures that were subjected to DNA expression array analysis. Shown is the percentage of cells that survived Ad5/35.IR-E1A/TRAIL (grey bars) treatment for 8 days or was GFP negative 24 hrs after Ad5/35.GFP (green bars) transduction. Standard deviation was less than 10%. **B)** Ad5/35.GFP infection (MOI 10 pfu/cell) of R/E and S/M cultures that were used in DNA expression arrays (N=31). Shown are GFP transduction rates and GFP mean fluorescence 24 hrs after infection. **C)** Microscopy of R/E and S/M cells 24 hrs after infection with Ad5/35.GFP (MOI 10 pfu/cell). **D)** Lowereing the cell density of R/E cells does not result in increased transduction rates. The level of GFP positive cells was quantified by flow cytometry 24 hrs after infection with Ad5/35.GFP (MOI 10 pfu/cell).

# Results

To understand the mechanisms of decreased attachment of Ad5/35 particles to epithelial cells, the levels of the primary Ad5/35 attachment receptor CD46, and of $a_V$-integrins (considered to be involved in uptake of adenoviruses targeting CD46 (Murakami et al., 2007)) were analyzed. Surprisingly, there was no significant difference both in the percentage of CD46 and $a_V$-integrin positive cells, and mean fluorescence levels between R/E clones, S/M clones and the initial (p1) ovc316m culture (Fig. 4.12A). Further investigation revealed that the architecture of epithelial cells mediates resistance to Ad5/35 attachment. Using confocal microscopy it was identified that the majority of CD46 was found inside tight junctions, with few receptor molecules localized to the apical membrane of R/E cells. In contrast, S/M cells had CD46 evenly distributed over the entire cell membrane. Cy-3 labeled adenovirus particles attached only to the apical side of R/E cells and not to CD46 trapped in tight junctions (Fig. 4.12B). On the contrary, in S/M cells, adenovirus particles attached to cells from both apical and lateral sides. When clusters of E/R cells were analyzed, tight junctions between cells excluded Cy3-Ad5/35 particles, whereas CD46 localized to the peripheral membranes of the clusters were accessible to virus binding in polarized cells. In R/E cells, $a_V$ integrins were found in tight junctions and on the basolateral membrane (Fig. 4.12C). Consequently, adenovirus particles that were attached to the few apically localized CD46 molecules cannot be internalized into R/E cells (Fig. 4.12D). In contrast, in S/M cells both CD46 and $a_V$-integrin co-localize, conferring efficient adenovirus internalization. The finding that CD46 and $a_V$-integrins are localized on two different membrane sites of R/E cells was corroborated by transduction studies in trans-well chambers, where Ad5/35.GFP vectors were applied either from the apical or basolateral sides of cells (Fig. 4.12E). While the percentage of GFP-expressing cells was comparable in S/M cells and ovc316m passage 1 cells infected from either, the apical or basolateral side, in R/E cells apical infection was markedly less efficient.

To clarify the observation that lowered cell densities in R/E cultures do not lead to increased levels of transduction (see Fig. 4.11D), isolated clusters of R/E cells were analyzed by confocal microscopy. Here, different morphologies could be observed for epithelial cells that are located either in the center or in the periphery of cell clusters. Whereas cells in the center maintained tight junctions to adjacent cells on their lateral membranes, cells in the periphery did not stain positive for tight junction protein claudin 7 on surfaces lacking neighboring cells. Peripheral cells rarely bound applied Ad5/35 vectors, most likely due to changes in polarization that resulted in extension of apical membranes to lateral sides, which were not adjacent to other cells. In contrast, cells that maintained tight junctions to several adjacent cells appeared to be polarized and could bind viral particles on lateral sides that were not sealed by tight junctions (Fig. 4.13).

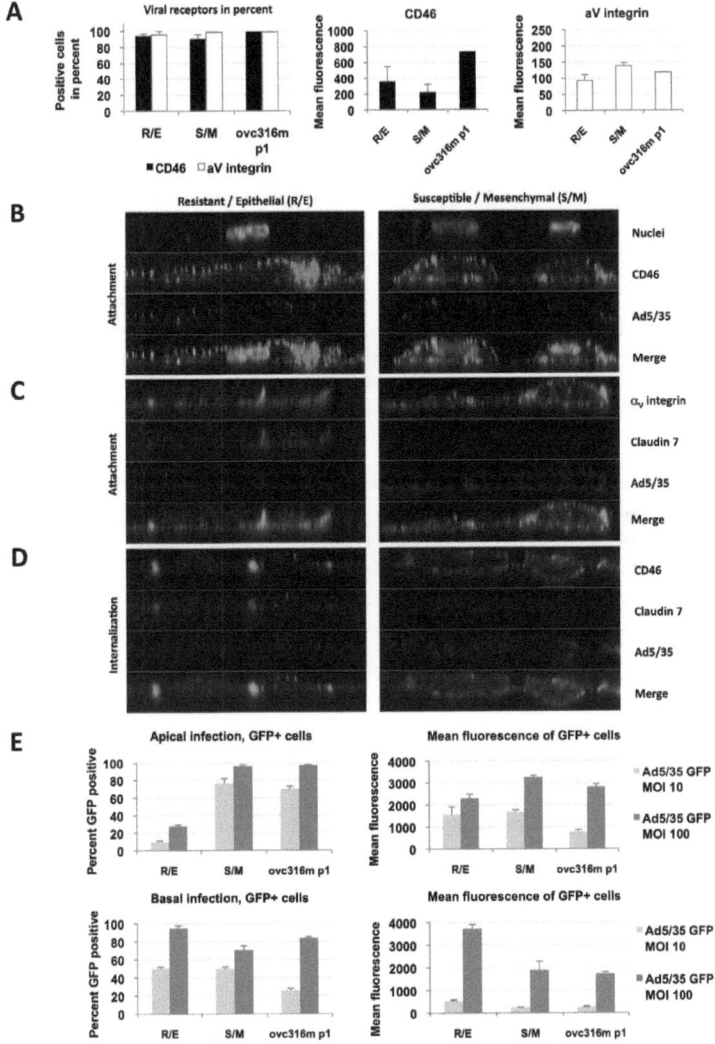

**Figure 4.12: Adenoviruses are excluded from membranes where viral receptors co-localize in R/E cells. A)** Flow cytometry analysis of CD46 and $\alpha_V$ integrins on R/E and S/M clonal cultures. Visualized are percentage of positive cells (left) and the mean fluorescence. **B-D)** Confocal microscopy, Z-axis visualized. R/E and S/M cells were incubated with 4000 Cy3-labeled Ad5/35 particles/cell for 30 min on ice (Attachment) or 2 hrs at 37°C (Internalization). Staining for CD46 and nuclei **(B)**, $\alpha_V$ integrin and claudin 7 **(C)**, and CD46 and claudin 7 **(D)**. **E)** Infection of cells with Ad5/35.GFP applied from apical and basal sides. R/E, S/M clonal cultures and ovc316m cells in passage 1 were seeded into tissue culture inserts in 48 well plates. Virus was applied in medium on top or below at an MOI of 10 or 100 pfu/cell. GFP expression was analyzed 48 hrs later by flow cytometry.

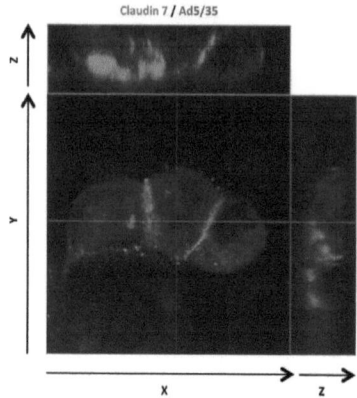

**Figure 4.13: Adenoviruses only infect polarized cells on exposed lateral membranes that are not sealed by tight junctions.** Confocal microscopy analysis of an epithelial cell cluster of R/E cells reveals different morphologies for cells within epithelial cell layers and cells located at the periphery of clusters. Intercellular junctions with adjacent cells polarize the central cell and virus is able to bind to the lateral sides that are not sealed by tight junctions. The two cells in the periphery, which maintain only tight junctions on one membrane, do not bind viral particles. Tight junctions are marked by claudin 7 (green), Ad5/35 particles appear in red.

## 4.4. The epithelial phenotype is a barrier to adenovirus infection *in vivo*

Next, it was tested whether the epithelial phenotype of cancer cells is also a barrier to Ad5/35 vector infection *in vivo*. First, tumor sections from ovarian cancer patients were stained for the epithelial adherens protein E-cadherin, and for laminin, a protein secreted by mesenchymal cancer cells (Fig. 4.14A). Large subsets of malignant epithelial cells appeared to be surrounded by tumor stroma, consisting of laminin in the primary tumor as well as in omentum and kidney metastases. This morphology could be reproduced in mouse xenografts derived from ovarian cancer cultures (Fig. 4.14A). In both patient tumors and xenografts, CD46 and $a_v$ integrins were co-localized with the tight junction protein claudin 7, supporting and validating the *in vitro* findings (Figs. 4.14B/C). This histology was observed for all biopsies from ovarian cancer patients as well as in xenografts derived from the ovarian cancer cell line SKOV3-ip1 (Fig. 4.15). As expected from this morphology, both intratumoral and intravenous injection of $2x10^9$ pfu of Ad5/35.IR-E1A/TRAIL into mice bearing subcutaneous ovc316 xenografts had no effect on tumor growth compared to PBS injected mice (data not shown). No viral replication (based on hexon staining) was detectable in ovc316 tumors at day 8 after intratumoral injection of Ad5/35.IR-E1A/TRAIL (Fig. 4.14D, left panel). Intratumoral injection of Ad5/35-GFP resulted in GFP expression only in cells directly surrounding the needle track (Fig. 4.14D, right panel). Intravenous injection of an Ad5/35 vector conferred transgene expression in sparse cells around the tumor periphery, a tumor area that appeared to be vascularized (Fig. 4.14E). Although the majority of cells in early passage ovc316 cultures were susceptible to Ad5/35 infection *in vitro* (see figure 4.1), *in vivo* transduction after Ad5/35 injection into ovc316 xenografts was very inefficient. To clarify this discrepancy, flow cytometry analyses for the epithelial marker EpCAM and the mesenchymal/mesenchymal-stem-cell markers vimentin and CD44 were performed on cell suspensions of ovc316 xenograft tumors and on cultured ovc316m cells from passage 1 and 20 (Figures 4.14F/G).

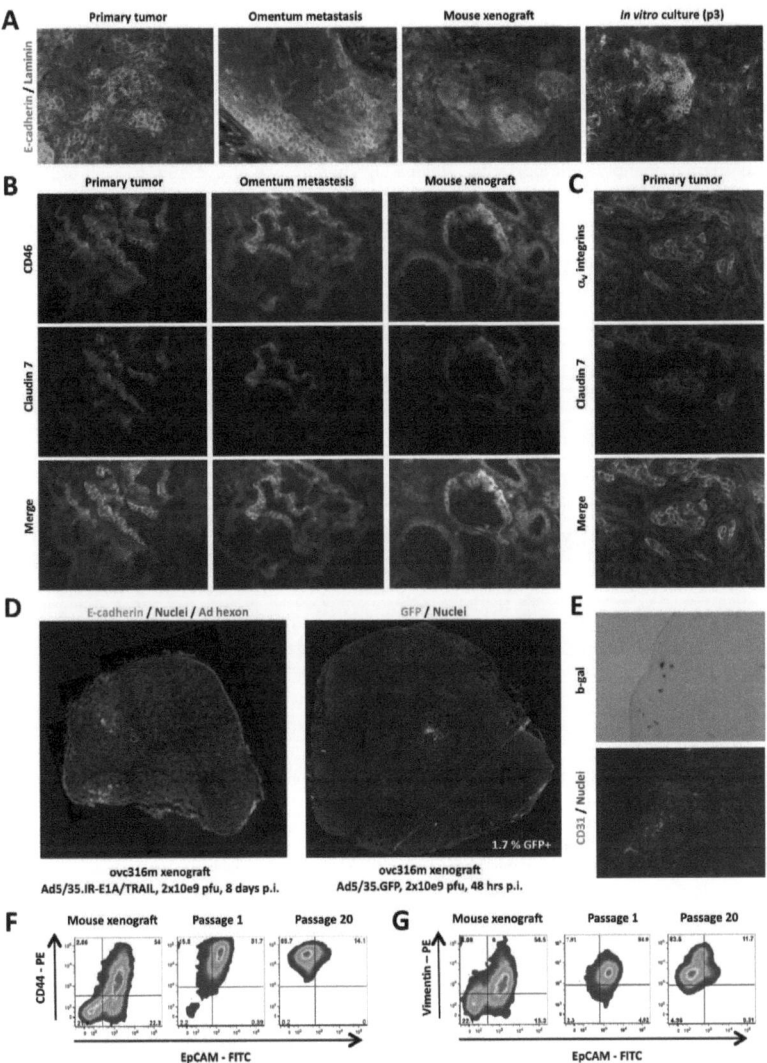

Figure 4.14: Ovc316 tumors have an epithelial phenotype and trap viral receptors in tight junctions. Shown are ovc316 tumor, matching metastases and xenografts (ovc316m). **A-E)** Nuclei are stained in blue. **A)** Analysis of tumor sections. E-cadherin (green) and laminin (red) expressing cells are existent in tumors and in *in vitro* culture (passage 3). **B)** Colocalization of CD46 (green) and tight junction protein claudin 7 (red). **C)** Colocalization of $\alpha_V$ integrins (green) and claudin 7 (red). **D)** Left panel: viral hexon (red) and E-cadherin (green) staining at day 8 post-injection of $2x10^9$ pfu of Ad5/35.IR-E1A/TRAIL. Right panel: *in vivo* GFP expression 48 hrs after injection of $2x10^9$ pfu Ad5/35.GFP. Percentage of GFP+ cells determined by flow cytometry. **E)** *In vivo* β-gal expression 48 hrs after intravenous injection of Ad5/35.β-gal. **F)** and **G)** Flow cytometry analysis of xenograft cell suspension, and cultured cells in passages 1 and 20. Staining for epithelial marker EpCAM and CD44 (F), EpCAM and mesenchymal marker vimentin (G). Cells in xenografts are either epithelial or in an epithelial/mesenchymal (E/M) hybrid stage. E/M hybrid cells adapt to tissue culture (passage 1) and differentiate into mesenchymal cells (passage 20).

In tumors, mesenchymal cells were rare (2% $CD44^{high}/EpCAM^{low}$ and 8% $vimentin^{high}/EpCAM^{low}$). The vast majority of ovarian cancer cells (>80%) expressed high levels of the epithelial marker EpCAM in vivo. Interestingly, almost all cells isolated from xenografts that adapted to tissue culture were positive for epithelial and mesenchymal markers. When further passaged, ovarian cancer cells lost EpCAM expression and differentiated towards the mesenchymal phenotype. Importantly, while a high percentage of ovc316m p1 cells are EpCAM positive by flow analysis, immunofluorescence studies revealed that most of the E-cadherin signal is either cytoplasmic or not co-localizing with lateral junctions, indicating that these cells are in the process of losing epithelial features, while gaining mesenchymal properties as N-p120 (see figure 4.7A).

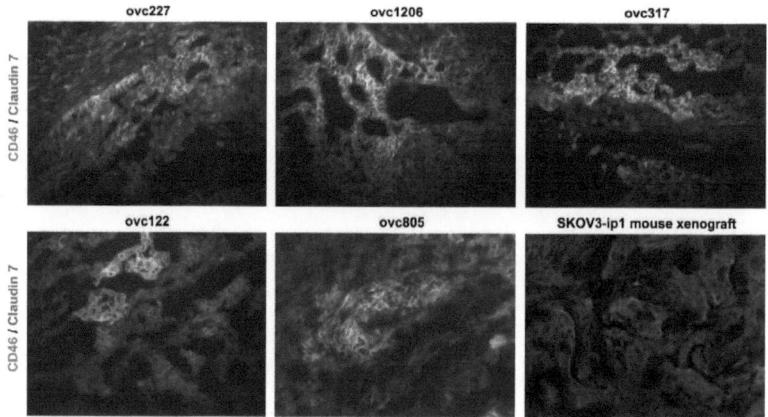

**Figure 4.15: CD46 is trapped in tight junctions of epithelial ovarian cancer cells *in situ*.** Adenovirus receptor CD46 (green) co-localizes with tight junction marker claudin 7 (red) in all tested sections of ovarian cancer patients and in xenografts derived from ovarian cancer cell line SKOV3-ip1. Nuclei are stained in blue.

In an attempt to generalize the finding that the epithelial phenotype is a barrier to adenovirus infection, other cell lines were tested for their phenotypes and correlating infectability with Ad5/35.GFP *in vitro* and in subcutaneous xenograft models (intratumoral injection). Analyzed were the epithelial ovarian cancer cell line SKOV3-ip1, HeLa cells (cervical adenocarcinoma), and the colon carcinoma cell line HT-29 (Fig. 4.16). Similar to what was observed for ovc316, there appeared to be a discrepancy between the phenotypes of ovarian cancer cell line SKOV3-ip1 cells *in vivo* and *in vitro*. While most of the cells in xenografts were epithelial, SKOV3-ip1 cells, subjected to culturing after isolation from xenografts, underwent an EMT. Interesting here was a difference in epithelial markers EpCAM and E-cadherin *in vitro*. Whereas the majority of cells maintained EpCAM on the cell surface, the presence of E-cadherin *in vitro* was low (Fig. 4.16A, left panel), indicating different degrees of EMT in primary cells and cell line.

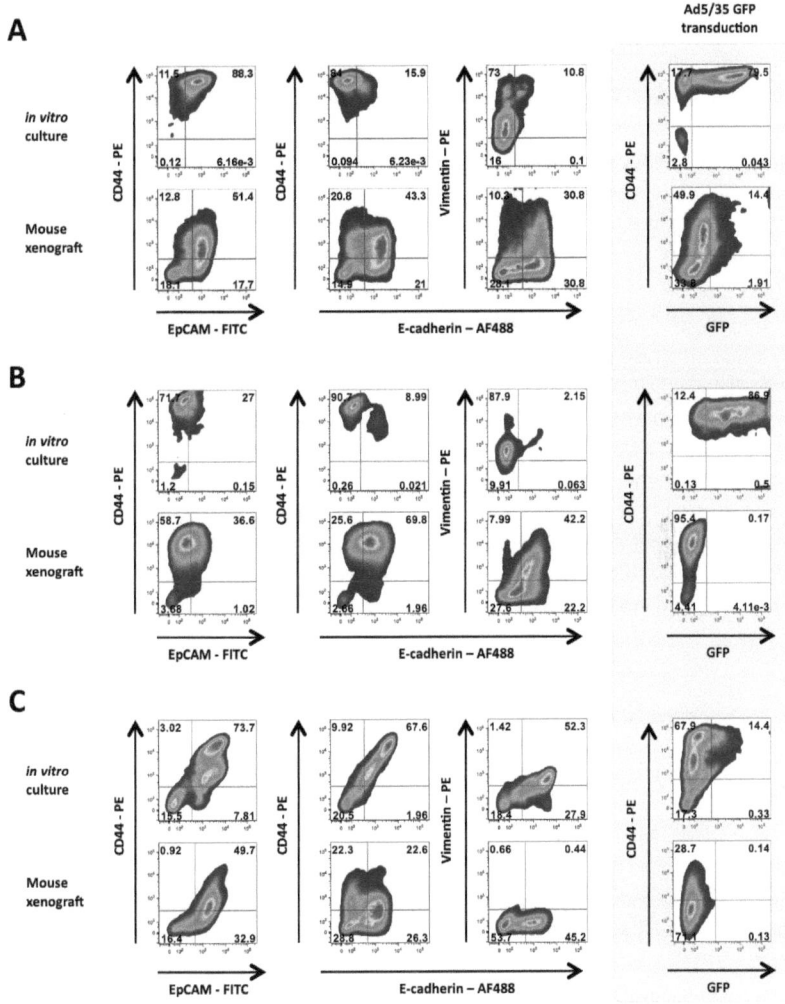

**Figure 4.16: Phenotypes of cells in *in vitro* cultures and subcutaneous xenografts differ greatly.** Top rows: *in vitro*, bottom rows: *in vivo*. Left panels: Stainings for epithelial markers EpCAM or E-cadherin on the X-axis, and mesenchymal markers CD44 or vimentin on the Y-axis. Right panels (grey box): GFP expression and CD44 intensity 48 hrs after infection with 100 pfu/cell (*in vitro*) or intratumoral injection of $2 \times 10^9$ pfu Ad5/35.GFP. **A)** SKOV3-ip1 cells show a discrepancy in epithelial markers EpCAM and E-cadherin in vitro and between phenotypes *in vivo* and *in vitro* (left panel). The discrepancy in phenotypes is reflected by the infectability and consequent GFP expression. **B)** HeLa cells express more E-cadherin than EpCAM *in vivo*. Almost all cells in xenografts are positive for CD44. The expression of vimentin is reduced *in vivo* (left panel) and no expression of GFP is detectable in tumors (right panel). **C)** HT-29 cells have epithelial phenotypes *in vitro* and *in vivo*. GFP transduction is low *in vitro* and absent in xenograft tumors.

Cultured SKOV3-ip1 cells were efficiently infectable with Ad5/35.GFP (80% GFP positive), while the epithelial xenograft tumor showed only low expression for GFP in about 14% of all cells after adenovirus infection (Fig4.16A, right panel). The number of mesenchymal cells in xenografts varied between 10 to 20%, depending on the markers vimentin and CD44 respectively. HeLa cells also showed a difference between their *in vivo* and *in vitro* phenotypes, correlating with their infectability. *In vitro* cultures were almost completely mesenchymal with only sparse epithelial cells. In contrast, HeLa xenografts had high levels of E-cadherin (about 70% of all cells) and showed reduced expression of vimentin, when compared to cultured cells (Fig.4.16B, left panel). No GFP positive cells were detected in xenografts after intratumoral injection of Ad5/35.GFP, whereas cells in tissue culture were >86% GFP positive after infection (Fig.4.16B, right panel). Interestingly, HT-29 cells did not undergo EMT *in vitro* and remained epithelial when cultured. Furthermore, the expression of mesenchymal marker vimentin was very low or absent *in vitro* and *in vivo* (Fig.4.16C, left panel). Subsequently, the infection with Ad5/35.GFP resulted in low GFP expression *in vitro* (about 14% of all cells) and absence of GFP *in vivo* (Fig.4.16C, right panel). In summary, poor Ad5/35 transduction rates correlate with the epithelial phenotype and the absence of mesenchymal marker vimentin. Among the tested epithelial cell line-derived tumors, SKOV3-ip1 xenografts showed the highest transduction rates in response to infection with GFP-expressing Ad5/35 vectors. Here, the numbers of transduced cells correlated with the fraction of non-epithelial vimentin$^{high}$ cells. This fraction was consistently low in all non-transducible xenografts.

## 4.5. The epithelial phenotype is also a barrier for adenovirus 5-based vectors

The finding that Ad5/35 receptors are trapped in tight junctions is reminiscent of the situation with the Ad5 receptor, CAR, which is known to be an integral tight junction protein (Cohen et al., 2001). Overall, CAR levels are significantly lower than CD46 levels on primary ovarian and cervical cancer cells and tumors *in situ* (personal conversation, Di Paolo and Lieber). Similar to CD46 and $a_V$-integrins, CAR levels were not significantly different between clonal R/E and S/M ovarian cancer cultures derived from ovc316 (Fig. 4.17A). Epithelial cultures could not be infected from the apical side with an Ad5 vector expressing GFP (Fig. 4.17B). Moreover, on sections of the ovc316 tumor and in ovc316m xenografts, CAR was found to be co-localized with the tight junction protein claudin 7 (Fig. 4.17C) and *in vivo* application of Ad5 vectors resulted in similarly inefficient tumor cell transduction as described above for Ad5/35 vectors (data not shown). Similar to what was seen for Ad5/35 based vectors, a correlation between resistance to viral oncolysis and the epithelial cell phenotype of ovarian cancer was observed by infection of culture ovc316m with an Ad5-based oncolytic vector (Ad5.IR-E1A/TRAIL). In immunofluorescence studies using E-cadherin/N-p120 (Fig. 4.18A) and E-cadherin/viral hexon (Fig. 4.18B) as markers, only cells positive for E-cadherin survived adenovirus treatment for 8 days. Flow cytometry for EpCAM and the MSC marker CD44 showed that the percentage of CD44$^{high}$/EpCAM$^{low}$ cells decreased, while the population of epithelial (E-cadherin$^{high}$) cells increased during days 2 and 8 post-infection with Ad5.IR-E1A/TRAIL (Fig. 4.18C-E). In summary, the epithelial phenotype of ovarian cancer, specifically the trapping of adenovirus receptors in tight junctions, represents a significant obstacle to transduction with both Ad5/35- and Ad5-based vectors.

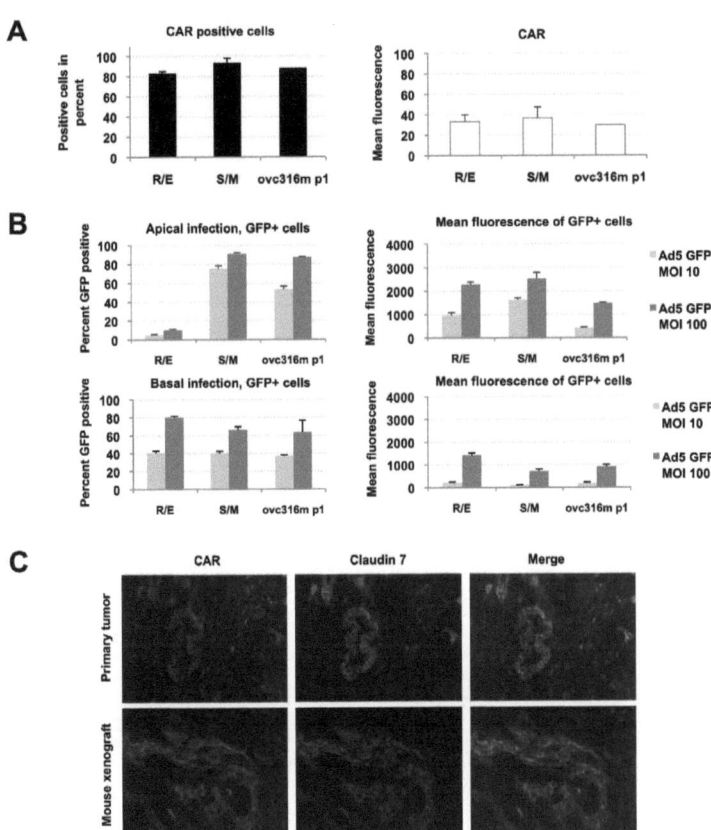

Figure 4.17: CAR-trapping in tight junctions translates to low apical transduction rates with Ad5. A) Flow cytometry analysis CAR on R/E and S/M clonal cultures. Visualized are percentage of positive cells (left panel) and the mean fluorescence (right panel). B) Infection of cells with Ad5.GFP applied from apical and basal sides. R/E, S/M clonal cultures and ovc316m cells in passage 1 were seeded into tissue culture inserts in 48 well plates. Virus was applied in medium on top or below at an MOI of 10 or 100 pfu/cell. GFP expression was analyzed 48 hrs later by flow cytometry. C) Colocalization of CAR (green) and tight junction protein claudin 7 (red) on cancer cells in primary ovc316 tumor section (upper panel) and ovc316m xenograft (lower panel). Nuclei are visualized in blue.

# RESULTS

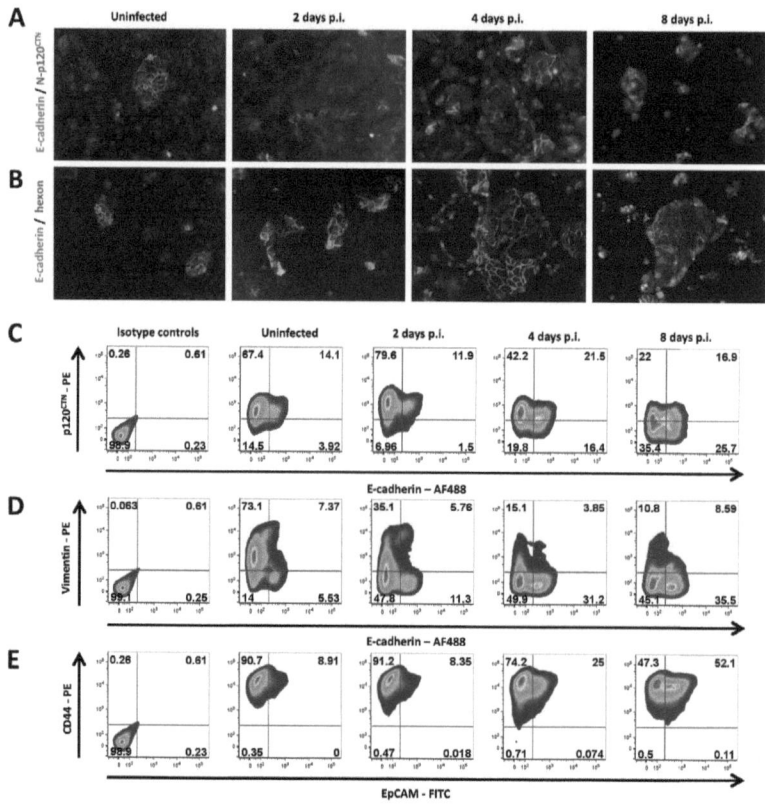

**Figure 4.18: Epithelial cells within culture ovc316m also survive treatment with oncolytic Ad5.IR-E1A/TRAIL.** A/B) Primary cultures (passage 10) were infected with Ad5.IR-E1A/TRAIL at an MOI of 100 pfu/cell and analyzed at days 2, 4, and 8 after infection by immunofluorescence staining for E-cadherin and N-p120 (A), E-cadherin and viral hexon (B). Nuclei are stained in blue. Uninfected cells were used as a control. C-E) Flow cytometry analysis of infected cells at days 2, 4, and 8 after infection shows enrichment for epithelial cells over time. Representative samples are shown.

## 4.6. Pathways involved in maintenance of the epithelial phenotype

It was hypothesized that transient opening of tight and/or adherens junctions would overcome resistance of ovarian cancer cells to Ad5 and Ad5/35 vector infection. As outlined above, almost all of the ovarian cancer cells in tumors were epithelial. Therefore, it was attempted to overcome resistance to viral oncolysis in R/E clones, which best resemble the *in vivo* histology. To understand regulation of tight and adherens junction pathways in ovarian cancer cells, the presence, activity or phosphorylation of key members of these pathways and regulators of the underlying F-actin cytoskeleton was studied in clonal R/E and S/M cultures, as well as in xenograft tumors and cultures derived from them at passage 1 and 20 by Western blotting. As expected, high levels of E-cadherin were found in R/E cells (Fig. 4.19A) and tumors, while with passaging of primary cells the level of E-cadherin decreased. In agreement with the immunofluorescence data (Fig. 4.7A), N-p120 is expressed at higher levels in S/M clones, and its expression increases with differentiation of epithelial/mesenchymal hybrid cells into mesenchymal cells during passaging of ovc316 cultures (Fig. 4.19A).

The observed differences in N-p120 levels between R/E and S/M clones indicate different impact of Rho GTPases in conferring resistance to Ad5/35 infection (see Fig. 4.4D) (Fox and Peifer, 2007). Along this line, different RhoA levels were detected in R/E and S/M clones. Also noteworthy is the higher amount of F-actin interacting Caldesmon in S/M cells and ovc316m cells that adapt to tissue culture. Caldesmon is involved in cytoskeletal remodeling and competes with the Arp2/3 complex, which crosslinks F-actin stress fibers. The differences in N-p120 levels in the tumor and passages 1 and 20 support earlier obtained immunfluoresence and flow cytometry data and demonstrate that cells, which adapt to tissue cultures, are different from tumor cells *in vivo* and that over passaging these cells differentiate towards mesenchymal cells. The Western blot analyses also indicate that R/E clones are closer to tumor cells *in vivo* and represent a more adequate model for attempts to overcome resistance. The analysis of ROCK, the downstream effector kinase of Rho, also revealed differences between R/E and S/M cells and tumor xenografts. Interestingly, ROCK expression was almost undetectable in R/E cells only. Like N-p120, active ROCK is reported to contribute to the invasive phenotype of epithelial cancers (Croft et al., 2004; Han et al., 2005), highlighting the importance of mesenchymal features for the susceptibility to viral infection and oncolysis. For focal adhesion kinase (FAK) a difference in between xenograft tumor cells and *in vitro* cultures was observed. Whereas all in vitro cultures predominantly contained the fully functional 125 kDa kinase, cells in xenografts harbored the proteolytically modified 90 and 40 kDa versions, which are known to be attenuated in their autokinase activity downstream of integrins (Cooray et al., 1996). To further analyze the impact of Rho GTPases on the resistance against viral internalization, R/E and S/M cells were infected with 200 pfu/cell for 15 min using plain medium without growth factors. The basal infection route was chosen in order to achieve comparable levels of viral infection on R/E and S/M cells (Fig. 4.19B). Control cells were treated with fresh medium (containing all supplements) that was used for all experiments in this study. There was higher activation of Rac1 after viral treatment in S/M cells, whereas only R/E cells showed activity for RhoA after infection. GTP carrying Cdc42 was not observed. Fresh medium induced activity of all GTPases, indicating the ability to remodel the cytoskeleton in response to growth factors. In R/E cells the p85 regulatory subunit of PI3K was phosphorylated at higher degrees after virus treatment and medium change, whereas p55 PI3K was specifically activated after virus treatment in R/E and S/M cells at comparable levels. To further validate the role of these pathways in maintenance of epithelial morphology and to potentially manipulate these pathways, a series of inhibitors was used, including exoenzyme C3 from *Clostridium botulinum* (inhibitor of Rho A, B and C GTPases), H-1152 (Rho-kinase inhibitor), *Clostridium difficile* toxin B (inhibitor for Rho, Rac, and Cdc42), Wortmanin (inhibitor of PI3K), as well as the Rac1 inhibitor.

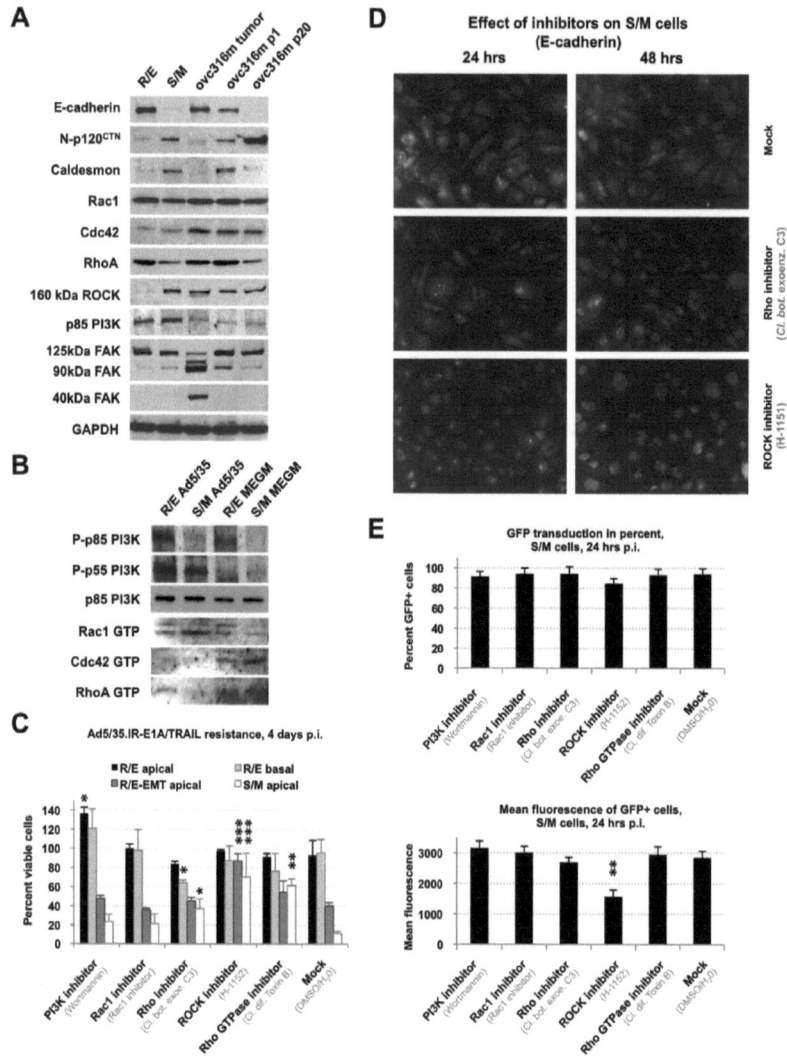

**Figure 4.19: Inhibition or lack of Rho kinase (ROCK) results in resistance to viral oncolysis. A)** Western blot for key members of pathways that regulate tight junction reorganization and EMT. **B)** Activity of PI3K and Rho GTPases in R/E and S/M clonal cultures 10 min after basal infection with Ad5/35 (200 pfu/cell) or treatment with fresh MEGM medium. **C)** MTT assay: Effect of inhibitors on viability of Ad5/35.IR-E1A/TRAIL infected R/E and S/M cells. Cells were pre-treated with inhibitors 24 hrs before infection. Rho and ROCK inhibitors enhance viral oncolysis in S/M cells. PI3K inhibition enhances cells survival of R/E cells (apical infection). Rho inhibition decreases amount of viable R/E cells after basal infection. **D)** Treatment of S/M cells with Rho or ROCK inhibitors for 24 or 48 hrs does not result in E-cadherin translocation to lateral membranes, but lead to a cobblestone-like morphology. **E)** Levels of GFP positive S/M cells and corresponding mean fluorescence 24 hrs after infection with Ad5/35.GFP (MOI 50 pfu/cell). The GFP transduction (mean fluorescence) is only significantly lowered in ROCK inhibitor-treated cells. *** $p<0.001$, ** $p<0.01$, * $p<0.05$

Treatment of S/M and R/E-EMT cells with Rac/Cdc42/RhoA, Rho (A/B/C) or ROCK inhibitors increased their resistance to killing by Ad5/35.IR-E1A/TRAIL, compared to cells that were treated with dilution buffer (Fig. 4.19C). This is most likely due to inhibition of Rho family GTPases that are involved in formation of tight/adherens junctions and that are also required for efficient integrin-mediated adenovirus internalization and/or intracellular trafficking. Surprising was the fact, that unlike observed earlier with GFP expressing viruses, basal infection of R/E cells had no enhancing effect on cell killing after 4 days compared to apical application of viral particles, indicating that viral replication and spread is reduced in R/E cells even after successful infection. However, additional inhibition of Rho could significantly increase cell killing on R/E cells. This suggests a dual function for Rho, which maintains the resistant phenotype of R/E cells independently of ROCK on one hand and increases the susceptibility of S/M cells in context with ROCK on the other. Supporting this hypothesis, Rho inhibition had no effect on a culture turning from R/E into S/M cells (R/E-EMT), while ROCK inhibition markedly increased cell survival. Inhibition of PI3K by Wortmanin increased the resistance of R/E cells compared to cells treated with drug-dilution buffer. Furthermore, the treatment with Rho GTPase or ROCK inhibitors had no effect on the distribution of E-cadherin in S/M cells, but resulted in a more cobblestone-like morphology that is characteristic for epithelial cells (Fig. 4.19D). Infection with a GFP expressing Ad5/35 vector resulted in high levels of transduction among inhibitor-treated S/M cultures. However, ROCK inhibition could significantly reduce the observed mean GFP fluorescence in S/M cells 24 hours after infection. Overall these data show that defects in the regulation of Rho GTPases or ROCK represent another independent mechanism that inhibits oncolysis in resistant cells in addition to up-regulated tight and adherens junctions and receptor trapping that prevents adenovirus infection.

In order to initiate infection, many pathogens must breach the epithelial barrier to gain access to the body. *Vibrio cholerae* strains produce zonula occludens toxin (ZOT), which is capable of reversibly altering intestinal epithelial tight junctions (Cox et al., 2002). After binding to its surface receptor, ZOT is internalized and subsequently triggers a series of intracellular events including phospholipase C and PKCα-dependent actin polymerization, which leads to opening of tight junctions (Fasano et al., 1995). However, the complete cascade of the intracellular events activated by ZOT is still elusive. Recombinant ZOT protein was produced in *E.coli* and the purified protein (4mg/ml) was applied to R/E cells for 2 hours (this time point is based on previous studies (Fasano et al., 1991). A disruption of F-actin fibers (Fig. 4.20A) and adherens junctions, formed by E-cadherin (Fig. 4.20B) was observed in ZOT treated R/E cells, compared to cells incubated with dilution buffer. Importantly, E-cadherin removal from the lateral membranes was restricted to individual foci within the tissue monolayer, whereas no differences in N-p120 intensities were observed (Fig. 4.20B). For transduction studies, ZOT-containing media was removed and cells were infected with Ad5/35.GFP. Transduction efficiency, based on the percentage of GFP-expressing cells, increased 1.7-fold as a result of ZOT treatment, but did not exceed 35% of R/E cells (Fig. 4.20C).

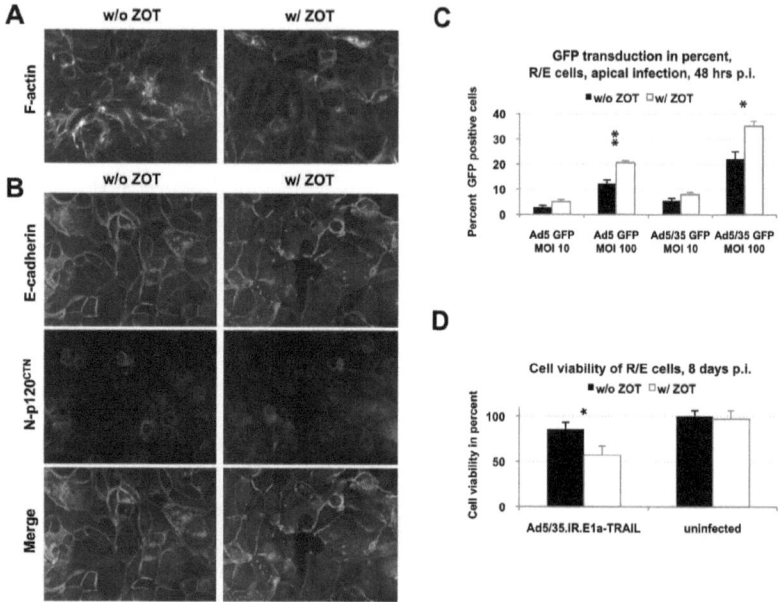

**Figure 4.20: Disruption of intercellular epithelial junctions by treatment with ZOT enhances adenovirus-mediated transduction and oncolysis.** R/E cells were used for all experiments. **A)** ZOT treatment (4µg/ml for 2 hrs) leads to F-actin relocation from lateral membranes throughout the cytoplasm. **B)** ZOT (4µg/ml for 2 hrs) induces a partly E-cadherin internalization and plaques within the tissue monolayer. **C)** ZOT treatment (4µg/ml for 2 hrs) increases viral transduction rates of Ad5 or Ad5/35.GFP (MOI 10 or 100 pfu/cell). GFP expression as analyzed by flow cytometry 24 hrs after infection (N=3). **D)** ZOT treatment (4µg/ml for 2 hrs) followed by infection with Ad5/35.IR-E1A/TRAIL at an MOI of 100 pfu/cell results in increased oncolysis. Cell viability was measured 8 days after infection by MTT assay (N=3). Controls were incubated with ZOT dilution buffer. ** $p<0.01$, * $p<0.05$

It was also analyzed whether ZOT pretreatment can overcome resistance of R/E cells to Ad5/35.IR-E1A/TRAIL. At day 8 after infection of R/E cells at an MOI of 100 pfu/cell, cytopathic effect was observed in clusters of Zot treated cells but not in dilution buffer treated cells. A quantitative analysis of cell viability using MTT assay revealed significantly less viable cells in ZOT pretreated cells that were infected with the oncolytic virus compared to control groups ($p<0.05$) (Fig. 4.20D). However, the percentage of lysed cells did not increase over time, indicating that repeated ZOT incubation is required to support viral spread. Notably, treatment of cells for longer periods (>2 hours) or with higher ZOT concentrations resulted in significant toxicity. Intratumoral injection of ZOT into subcutaneous ovc316 tumors appeared to result in re-localization of F-actin and claudin 7 within 2 hours of injection, but also caused death of treated animals, preventing further studies with adenoviruses. In summary, while ZOT appears to transiently rearrange the cytoskeleton and support infection of resistant ovarian cells, this approach is relatively inefficient, requires repeated application, and is associated with toxicity.

A different attempt to disrupt intercellular epithelial junctions of R/E cells is the depletion of calcium. Maintenance as well as initiation of adherens and tight junctions is calcium-dependent (Rothen-Rutishauser et al., 2002). It was therefore tested, whether calcium depletion by versene (0.2 g/L EDTA in PBS) prior to infection could enhance viral oncolysis. Versene treatment for 3 hours induced the detachment of multiple cells, whereas the majority of R/E cells remained attached and started to clusterize. Immunohistochemistry analysis could confirm that tight junction protein claudin 7 was delocalized to the cytoplasm and tissue integrity was disrupted for a large number of cells in response to versene treatment (Fig. 4.21A). Versene pre-treated R/E cells were then infected apically with Ad5.IR-E1A/TRAIL or Ad5/35.IR-E1A/TRAIL in an MOI of 100 pfu/cell. Detached cells were collected and re-applied to according wells before infection. 8 days after infection, enhanced oncolysis was observed in versene-treated wells when compared to non-treated cultures by MTT assay (Fig. 4.21B and C). However, about one third of cells in virus-infected cultures remained viable and were primarily found in cell clusters. Versene treatment alone also lowered the number of viable cells in uninfected wells. This was most likely due to the initial cell detachment and redistribution, since no signs of cytotoxicity were observed. In summary, versene pre-treatment of R/E cells causes internalization of tight junction protein claudin 7 and significantly increases adenovirus-mediated oncolysis.

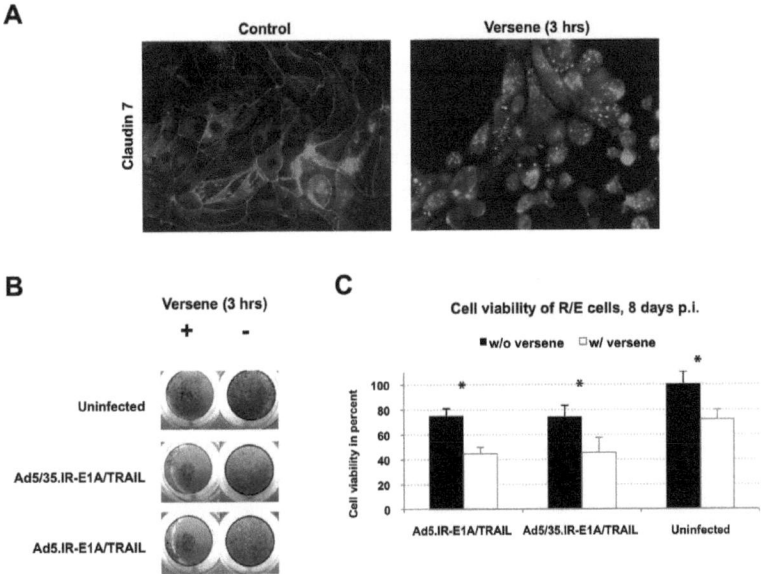

Figure 4.21: Calcium depletion induces Claudin 7 internalization and enhances viral oncolysis. A) Claudin 7 is internalized upon calcium depletion by versene for 3 hrs. Note that a number of cells detached from the plate. B) R/E cells were treated with versene for 3 hrs and then infected with indicated oncolytic adenoviruses. Shown are cultures before formazan crystals were dissolved for MTT assay readout. Large clusters of epithelial cells survive oncolytic virus treatment. C) Versene pre-treatment significantly reduces the amount of viable cells in virus-infected cultures and also decreases the initial number of cells in this experiment. MTT assay of triplicates, Infected cultures are scaled to appropriate controls. * p<0.05

## 4.7. Adenoviruses that target receptor X trigger E-cadherin removal

Human species B adenoviruses can be divided in 3 groups based on their receptor usage (Tuve et al., 2006). Group 1: (Ad16, 21, 35, 50) nearly exclusively utilize CD46 as a receptor; Group 2: (Ad3, Ad7, 14) share the same un-identified receptor/s which is currently refered to as receptor X; and Group 3: (Ad11p) preferentially interacts with CD46, but also utilizes receptor X if CD46 is blocked (Tuve et al., 2006). Recently, it was found that the Ad3 fiber knob interacts at low affinity with cellular heparan sulfate proteoglycans (HSPGs), including syndecan-4 (Tuve et al., 2008). In this study it was suggested that Ad3 knob binding to HSPGs subsequently increases high affinity interaction between the Ad3 virion and receptor X, whereby this interaction involves a viral capsid protein other than the fiber knob. This dual hit mechanism of infection is reminiscent of coxsackie B virus infection which involves initial binding to a primary receptor (decay-accelerating factor) thus triggering intracellular signals that permit the virus to move to tight junctions, interact with CAR, and enter the cell (Coyne and Bergelson, 2006). Based on this, it was hypothesized that Ad3, Ad7, and Ad14 were better candidates for achieving infection of R/E cells than Ad5 and Ad35, serotypes that initially bind to CAR and CD46, respectively. First, the presence of HSPGs on the membrane of R/E cells was analyzed using an antibody specific to N-sulfated glucosamine residues within heparin sulfate. Importantly, HSPGs were not strictly localized to lateral junctions only and showed a more ubiquitous distribution in multiple cells (Fig. 4.22A). Next, the binding/uptake of Cy-3 labeled Ad3 and Ad35 was studied on R/E cells. When cells were incubated for one hour with virus, Cy-Ad3 particles were found on most R/E cells, whereas cell-associated Cy-Ad35 signals were detectable on only sparse R/E cells (Fig. 4.22B). Incubation of R/E cells with adenovirus virions mediated removal of E-cadherin from the cell surface for Ad3, 7, 11, and 14 but not for the CAR-interacting Ad5 and the CD46 interacting Ad35 (Fig. 4.22C) and Ad5/35 (data not shown). This study also shows that Ad3, 7, 11, and 14 are able to kill R/E cells, which is reflected in plaque-like foci within the cell monolayer with hexon expressing cells along the periphery of the lysis plaques. Changes of membrane E-cadherin in R/E cells depending on the serotype used for infection are corroborated by flow cytometry studies (Fig. 4.22D). Mean E-cadherin fluorescence on R/E cells infected with Ad3, 7, 14, and Ad11 was about one order of magnitude less than in Ad5, Ad35, and Ad5/35 infected cells. Interestingly, levels of the epithelial cell marker EpCAM did not change after infection with different adenovirus serotypes, underscoring the role of E-cadherin in determining whether cells are susceptible to adenovirus infection or not. Notably, because CD46 is not accessible in R/E cells, Ad11 uses the alternative HSPG/receptor X pathway and therefore groups together with Ad3, 7 and 14. To further corroborate this finding, R/E cell killing by adenovirus serotypes was measured by MTT assay. This also revealed a clear grouping of tested Ad serotypes (Fig. 4.22E). Ad3, 7, 11, and 14 were significantly more efficient in lysing R/E cells than Ad5 and Ad35 ($p<0.001$). Next, it was attempted to confirm this superiority of one of these seroptypes, Ad3, *in vivo*. After intratumoral injection into subcutaneous ovc316 xenograft tumors, markedly more hexon positive cells were found in Ad3-injected tumors than in Ad35-injected tumors (Fig.4.22F).

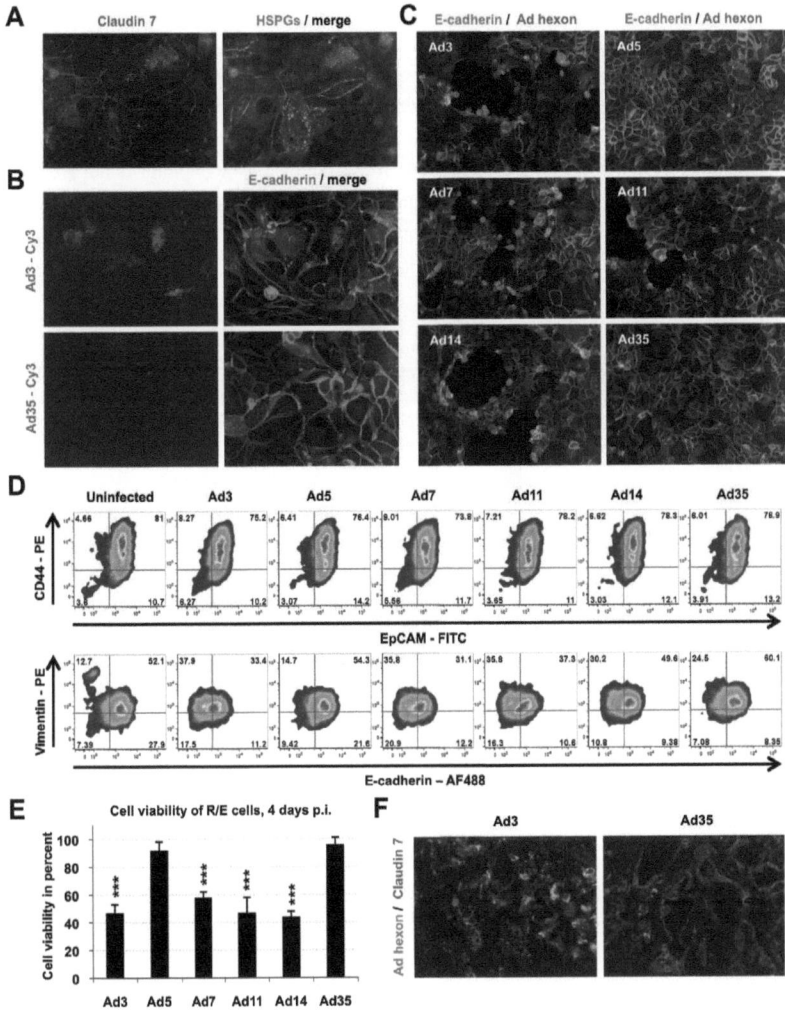

**Figure 4.22: Wild-type adenoviruses targeted to receptor X have superior infectious abilities on epithelial tissues.**
**A)** Expression of HSPGs (green) and claudin 7 (red) on R/E cells. **B)** Cy3-Ad3 and Cy3-Ad35 attachment (left panel) and internalization (right panel) in R/E cells. **C)** Effect of Ad 3, 5, 7, 11, 14, and 35 infection (MOI 100pfu/cell) on cell morphology and E-cadherin (green) expression in R/E cells. Viral replication was visualized by staining for hexon (red). Analysis was carried out at day 4 post infection. **D)** Flow cytometry analyses of Ad infected cells (same conditions as in C) reveal removal of E-cadherin after treatment with adenoviruses targeted to receptor X. In contrast, levels of EpCAM remain unchanged. **E)** Viability of R/E cells infected with wtAd serotypes at an MOI of 25pfu/cell. Viability was measured by MTT assay 4 days later (N=3). **F)** In vivo transduction of wAd3 and wtAd35. A total of $2 \times 10^9$ pfu of Ad3 and Ad35 was intratumorally injected into subcutaneous ovc316 tumors and tumor sections were analyzed 3 days later for claudin 7 (red) and Ad hexon (green) using an anti-hexon antibody that cross-reacts with all Ad serotypes. Note that more hexon positive cells were observed in tumors injected with Ad3. *** $p<0.001$

## Results

To quantitatively assess *in vivo* transduction, RNA was isolated from virus-injected tumors and hexon-mRNA levels were analyzed. Hexon mRNA levels of Ad3-injected tumors were about 10-fold higher compared with tumors that received Ad35 at day 3 post-injection, and continued increasing by day 11, indicating viral replication (Fig. 4.23A). Furthermore, it was observed that a single administration of $2 \times 10^9$ pfu of wtAd3 could delay tumor growth, whereas wtAd35 injection had no therapeutic effect (Fig. 4.23B). It was not possible to monitor tumor growth for longer periods, because mice became moribund at day 10 after injection. This was attributed to low-level replication of wildtype virus in normal tissue. At necropsy, enlarged livers and spleens were observed in virus-injected animals.

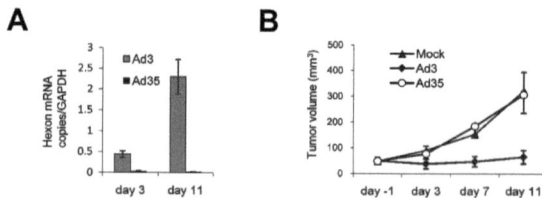

**Figure 4.23: Receptor X-targeted adenovirus serotype 3 shows superior oncolytic abilities in ovc316 xenografts.** A total of $2 \times 10^9$ pfu of wtAd3 or wtAd35 was injected intratumorally into subcutaneous ovc316 tumors. **A)** Transduction was determined by quantitative reverse transcription-PCR for hexon mRNA using pan-serotype hexon primers that can detect both Ad3 and Ad35 hexon mRNA. Hexon expression at day 3 and 11 post injection of wtAd3 or wtAd35 into xenografts is standardized to the level of GAPDH mRNA (N=5 each). **B)** Shown is the ovc316m tumor volume after mock injection or injection of $2 \times 10^9$ pfu of wtAd3 or wtAd35 (N=5 each). Intratumoral Ad3 injection could delay tumor growth of ovc316 xenografts.

# 5. Discussion

Oncolytic adenoviruses have demonstrated great potential as anti-cancer therapeutics in pre-clinical applications. However, their evaluation in clinical trials showed an apparent inability to spread throughout the tumor tissue and induce efficient oncolysis in solid tumors. The aim of the present thesis was the identification of cellular attributes and/or mechanisms that confer resistance to targeted virotherapy with adenoviruses. In this study, three different phenotypes of ovarian cancer cells were discovered: epithelial, mesenchymal, and hybrid. Using gene expression profiling of primary human ovarian cancer cells either resistant or susceptible to viral oncolysis, it was identified that the epithelial phenotype of cancer cells represents a barrier to infection by oncolytic adenoviruses targeted to CAR, CD46, or $a_V$-integrins. Specifically, it was showed that these adenovirus receptors are trapped in epithelial tight junctions and therefore not accessible to virus binding. Furthermore, it was demonstrated that even after successful infection, adenovirus-mediated oncolysis is impaired in epithelial-restricted cells. Interestingly, xenograft tumors derived from primary ovarian cancer culture ovc316 appeared to be exclusively epithelial with a high number of cells in an epithelial/mesenchymal (E/M) hybrid stage. E/M cells were the only cells that were able to adapt to tissue culture conditions. These cells rapidly lost epithelial markers as EpCAM and E-cadherin and generated a culture of mesenchymal cells when passaged *in vitro*. Importantly, E/M cells that already acquired mesenchymal features *in vitro* and mesenchymal cells represent the cell populations that are susceptible to commonly used oncolytic adenoviral vectors. Finally, it was demonstrated that adenoviruses, that utilize receptors other than CD46 or CAR, could efficiently lyse epithelial-restricted cells *in vitro* and *in vivo*. The significance of these findings obtained during the course of this study, their incorporation into currently existing models of barriers to virotherapy, and the perspectives they raise are discussed below.

## 5.1. Establishment and characterization of *in vitro* cultures that are resistant to viral oncolysis

When this study was initiated, *in vitro* resistance to adenoviral oncolysis was rarely observed. Studies analyzing the performance of tumor-targeted adenoviruses were usually carried out on cell lines or primary cancer cells and resistance was mostly defined by time delays in efficient cell-killing or seen in correlation with levels of adenoviral receptors. It was therefore critical to establish a new *in vitro* system, which could reflect the resistance to adenoviruses that is associated with solid tumors. Considering that the genetic instability of cancer cells leads to a high heterogeneity within human tumors (Loeb et al., 2008), it was thought that the employment of primary ovarian cancer cultures would be advantageous for the isolation of cells, which could survive virotherapy. This hypothesis could be validated for about 10% of all tested cultures, which contained subsets of cells that were able to escape adenoviral oncolysis even when high doses of virus were used. However, the vast majority of cells within cultures that contained resistant subsets were susceptible to Ad5/35.IR-E1A/TRAIL. In order to identify

## DISCUSSION

mechanisms that underlie the observed resistance, it was planned to compare cells that are either resistant or susceptible to viral oncolysis by DNA expression arrays. However, the extremely low frequency of surviving cells within primary cultures made this initial approach unsuitable for direct analysis. To exclude gene expression changes, which most certainly would be triggered due to the viral infection process, it was crucial that the analysis of cells resistant to oncolysis would mirror the situation before viruses were applied. In order to allow conclusive and unbiased readouts based on the mRNA level of uninfected cultures, the majority of cells in a given culture must therefore critically be resistant to oncolysis. Interestingly, primary cultures that contained resistant cell subsets showed a high morphologic heterogeneity. Among these cultures, ovc316 also had tumor forming properties, which was essential in order to link the yet to be identified *in vitro* resistance mechanisms to the situation in tumor xenografts derived from the same cells. Furthermore, primary cultures generated from patient biopsies generally showed contamination with fibroblasts, which additionally complicated direct readouts. As a positive side effect of cultures isolated from ovc316 xenografts (termed ovc316m) these fibroblasts were eliminated. Therefore, culture ovc316m was chosen to establish clonal cultures that were propagated and then tested for resistance to viral oncolysis. The analysis here concentrated on the CD46-targeted Ad5/35.IR-E1A/TRAIL because of the superior performance in the initial experiment on primary cell cultures. In this context it is noteworthy, that the enhanced oncolytic activity of this vector over an otherwise similar oncolytic adenovirus that uses CAR as a primary attachment receptor (Ad5.IR-E1A/TRAIL) was not surprising, because Ad5-based vectors carrying fiber proteins of adenovirus serotype 35 in their capsids have been shown to be more efficient in transduction and oncolysis in multiple studies before (Hoffmann et al., 2008; Sova et al., 2004). However, the screening of 100 clonal ovc316m cultures ultimately revealed that it is possible to establish cell lines, which are either completely resistant or susceptible to viral oncolysis from primary cultures that contain resistant cell subsets. Notably, the number of clonal cultures resistant or susceptible to adenovirus-mediated oncolysis was similar (20% and 19% respectively), while the majority of cultures showed heterogeneous responses to Ad5/35.IR-E1A/TRAIL.

### 5.2. The epithelial phenotype as a barrier for adenovirus infection and oncolysis

Because all of these cultures were derived from single cells, it was evident that distinct cells within culture ovc316m had abilities to generate both resistant and susceptible populations. DNA expression array analysis and the subsequent use of gene ontology software could reveal that the resistance of clonal ovc316m cultures to Ad5/35.IR-E1A/TRAIL correlates with the expression of multiple epithelial junction proteins. Further analysis identified the maintenance of the epithelial phenotype as a cellular mechanism conferring resistance to adenoviral infection and oncolysis. Access to viral receptors was critically linked to depolarization and the loss of tight and adherens junctions, both hallmarks of EMT and the subsequent mesenchymal phenotype.

## DISCUSSION

### 5.2.1. Epithelial tight junctions as a barrier for adenoviral infection

An apical-basal polarization in resistant/epithelial (R/E) cells led to a discrepancy in the distribution of viral receptors on distinct membranes. The primary attachment receptor CD46 was found primarily on lateral membranes with a few receptor molecules sorted to the apical surface on polarized R/E cells. Access to receptors in the paracellular space between adjacent cells was inhibited for Ad5/35 particles in R/E cells and successful binding to apically located CD46 did not translate into internalization of viral particles. This implies that co-receptors necessary for successful infection are not present on apical membranes. The uptake of adenoviruses targeting CAR is dependent on intercactions of $\alpha_v$integrins with viral pentons (Nemerow and Stewart, 1999; Wickham et al., 1993) and a recent report suggests similar processes for CD46-targeted Ad35 (Murakami et al., 2007). Interestingly, in monolayers of R/E cells $\alpha_v$integrins were exclusively sorted to basolateral membranes and therefore only co-loacalized with CD46 in the paracellular space. An apparent mismatch in the distribution of primary attachment receptor CD46 and $\alpha_v$integrins on polarized R/E cells can therefore be seen as a major factor for the poor transduction rate of apically applied Ad5/35 vectors. The importance of tight junctions, which exclude viral particles from the paracellular space, was further highlighted by the observation that apically applied Ad5/35 vectors almost exclusively transduced R/E cells in contact with the rim of tissue culture plates or located on top of monolayers (see Fig. 4.11C). Both of these areas inhibit cells in establishing tight junctions due to the lack of adjacent cells. Nevertheless, experiments in trans-well chambers showed that it is also possible to efficiently infect and transduce monolayers of R/E cells when adenoviral vectors were applied from the basal side. Here, the circumvention of epithelial tight junctions gave access to basal $\alpha_v$integrins and probably to CD46 in paracellular areas located underneath tight junctions. It is also possible that access to $\alpha_v$integrins alone might have mediated efficient virus entry using this infection route. However, earlier studies have shown that fiber-independent infection processes are inefficient on epithelial cells (Von Seggern et al., 1999). The exclusion of viral particles from the paracellular space is not the only mechanism how R/E cells can prevent adenovirus-access to viral receptors. Confocal microscopy analysis also suggested that cells in the periphery of isolated epithelial clusters might undergo changes in polarization, resulting in apical-like membranes on lateral sides. The binding of adenoviral particles would therefore be exclusive to cells that maintain a regular apical-basal polarity and expose lateral membranes, which are not sealed by tight junctions (Fig. 4.13). As observed during this study (Fig. 4.11D), lowered cell densities of polarized epithelial cells can therefore not be translated into increased adenovirus-mediated transduction rates. This mechanism might also explain why epithelial cell clusters survived treatment with Ad5/35.IR-E1A/TRAIL on low passage ovc316m cultures (Fig. 4.8). Here, similar conclusions could be drawn for an analogous vector that is targeted to CAR (Ad5.IR-E1A/TRAIL, Fig. 4.18) and CD46 targeted Ad5/35Δ24Ki/Cox (Fig. 4.10), which uses a different cell killing mechanism. These results imply a general resistance mechanism of epithelial restricted cells derived from ovc316 to CD46 or CAR targeted adenoviruses.

# DISCUSSION

These new observations showing adenovirus interactions with epithelial cancer cells can be brought in context with several earlier reports using CAR targeted Ad2 and Ad5 on human airway epithelial cells. It was known before that CAR is an integral tight junction protein (Cohen et al., 2001) and the apical Ad5 infection efficiency was limited on polarized cells (Walters et al., 1999). Like other tight junction members (e.g. claudins), CAR forms transcellular homodimers between neighboring cells (van Raaij et al., 2000) and overexpression of CAR leads to increased transepithelial resistance (Cohen et al., 2001). The fiber of Ad2 shows a higher affinity to CAR than CAR itself and triggers CAR homodimer disruption (Freimuth et al., 1999). Walters et al. found that upon initial infection of polarized airway epithelial layers, the first progeny virus was released to the basolateral surfaces and then traveled through the paracellular space towards the apical side (Walters et al., 2002). An excessive production of soluble fiber protein and the release of defective viral particles resulted in disruption of CAR homodimers and a concomitant decrease in transepithelial resistance. This was also accompanied by a breakdown of tight junctions in general, which ultimately released functional viral particles to the apical surface. While the authors interpreted this as a consequent effect of the CAR-CAR disruption, they also stated that they could not exclude that fiber binding to CAR triggers an intracellular signaling that contributes to the increased epithelial permeability. They also left the possibility that binding of viral pentons to integrins might facilitate adenovirus movement across epithelial barriers. Furthermore, it was left unclear how Ad2 can induce an efficient initial infection in polarized airway epithelial layers (Goosney and Nemerow, 2003). Importantly, the loss of tight junctions is a hallmark of EMT and the subsequent acquisition of the mesenchymal phenotype, which would render cells susceptible to adenovirus infection and oncolysis.

Interesting in this context is the finding, that ovc316 xenograft tumors almost exclusively contain cells in an epithelial or E/M hybrid stage based on their markers (Fig. 4.14F/G). Cells that adapted to tissue culture were in an E/M hybrid stage and almost indistinguishable from R/E cells by flow cytometry analysis (Fig. 4.5C). Further analysis revealed that differences between cultured E/M and R/E cells lay in intracellular processes that enforce EMT, depolarization and subsequent access to viral receptors. E/M cells that were isolated from xenograft tumors (ovc316m p1) mostly showed an irregular distribution of E-cadherin, which indicates the absence of an apical-basal polarity (Fig. 4.7A). Subsequently, infection of these cells with GFP expressing adenoviruses resulted in efficient transduction. In sections of primary ovarian carcinomas and xenografts, epithelial tumor cells nests stained positive for E-cadherin and claudin 7 on all membranes adjacent to other epithelial cells (Fig. 4.14). This suggests that multilayers of epithelial ovarian cancer cells *in vivo* maintain tight and adherens junctions on all surfaces. The conversion from this 3-dimensional complexity of intercellular junctions within solid tumors to a planar monolayer culture was probably largely responsible for the observed disorganized E-cadherin abundance in E/M cells that just adapted to plastic ware. However, over passaging the majority of E/M cells lost epithelial features and turned into mesenchymal cells, thereby generating a culture with artificial character when compared to xenografts. This implies that the cell biology of populations of primary ovarian cancer cell cultures and cancer cells *in situ* is different and that population cell cultures have only limited value for studying resistance to adenovirus infection. The discrepancy between *in vivo* and *in vitro* phenotypes was not only observed for primary culture ovc316 and could be reproduced in other cell lines, suggesting a more widespread phenomenon. This finding potentially explains why studies with established cells lines in the past have not revealed that the epithelial phenotype of cancer cells represents a barrier to adenovirus infection (Behzad et al., 2006; Davison et al., 2001; Kirby et al., 2004; Maxwell and Davis, 2000). Although differences between tissue integrity in *in vivo* and *in vitro* cultures were observed before, the reduced ability of adenoviruses to infect E-cadherin positive cells in culture, was entirely linked to low levels of CAR in these cells (Anders et al., 2003).

### 5.2.2. The epithelial phenotype as a barrier for adenovirus-mediated oncolysis

Noteably, using tumor cell oncolysis as the endpoint for resistance studies, DNA array expression analysis revealed the significantly altered expression of 983 genes. Only 33 of these were involved in tight junctions and cell adhesion pathways, which indicates that levels other than receptor trapping in tight junctions are present in epithelial-restricted ovc316 cells that confer resistance to oncolytic adenoviruses.

Another interesting difference found during the course of this thesis was the elevated expression and accumulation of multiple extracellular matrix (ECM) components by susceptible/ mesenchymal (S/M) cells (Fig 4.6A). Importantly, the ECM can trigger cell growth via signaling through integrins and focal adhesion kinase (FAK) (Pirone et al., 2006). Accordingly, S/M cells showed higher proliferation rates than R/E cells. In contrast, R/E cultures expressed the laminin-related Netrin-4, which can disrupt extracellular laminin networks (Schneiders et al., 2007) and inhibit angiogenesis in solid tumors (Lejmi et al., 2008). The enhanced proliferation rate of S/M cells was most likely another reason for the rapid increase of mesenchymal cells in ovc316 cultures during passaging (Fig. 4.14F/G). However, the clonal expansion of ovc316 cells shows that it is possible to isolate cultures that are completely epithelial and maintain apical-basal polarity, which renders them resistant to adenovirus infection and oncolysis. It was also observed that a number of R/E cultures were able to partly turn into S/M cultures during further passaging by undergoing an EMT (R/E-EMT cells). Importantly, other R/E cultures always maintained the epithelial phenotype when passaged in the same medium that induced EMT in the majority of ovc316m cells and R/E-EMT cells. Despite their discrepancy in the ability to generate or partly turn into S/M cells, R/E and R/E-EMT cultures were restricted to the epithelial phenotype and therefore resistant to viral oncolysis at point of initial testing. Importantly, EMT was inhibited for multiple clonal R/E cultures and mesenchymal features as N-p120 were consistently low in these cells. The mesenchymal version of this catenin was specifically monitored, because of its known regulating effect on members of the Rho family of GTPases. These small G proteins are major regulators of the actin cytoskeleton and additionally control a wide variety of molecular processes, including cell polarity, microtubule dynamics, vesicle trafficking, and transcription factor activity (Vega and Ridley, 2008). The Rho family comprises 20 proteins of which the best-studied members are RhoA, Rac1 and Cdc42. The impact of these three GTPases on adenovirus uptake was studied in A549 cells before. Viral internalization was

# DISCUSSION

dependent on active Rac1 and/or Cdc42 after signaling of Phosphoinositide3-kinase (PI3K) and focal adhesion kinase (FAK) via adenovirus attachment to $a_V$-integrins (Li et al., 1998a; Li et al., 1998b). Supporting this, PI3K activation could also be detected in R/E and S/M cells after adenovirus infection from the basal side. However, this infection route resulted in different responses in the activity of RhoA and Rac1 (Fig. 4.19B), despite high transduction rates in both cultures (Fig. 4.12E). Higher Rac1 activity was observed for S/M cells, whereas R/E cells responded with RhoA activity to infection with adenovirus. Even though these data are not fully conclusive, they suggest that adenovirus internalization processes differ between R/E and S/M cells. Furthermore, RhoA levels were elevated in E-cadherin positive cultures (Fig. 4.19A) and cell numbers could be significantly reduced after basal Ad5/35.IR-E1A/TRAIL infection in Rho-inhibited R/E cells (Fig. 4.19C). Conversely, N-p120, a known inhibitor of adherens junction-localized RhoA (Yanagisawa et al., 2008), was specifically prominent in all cultures that were susceptible to adenoviral oncolysis and underwent EMT. This strongly suggests that RhoA has a positive effect on the maintenance of the epithelial phenotype and subsequent cell survival in response to oncolytic adenoviruses. A potential mediator of the epithelial phenotype is Dia1. This RhoA effector protein was found significantly upregulated in R/E cultures (Table 4.1) and is known to ensure a dynamically stable interface between cells by sustaining adherens junction complexes (Sahai and Marshall, 2002).

As mentioned in paragraph 4.6, RhoA appears to have another function in context with its downstream effector kinase ROCK. Specifically, the inability of R/E cultures to undergo EMT might be directly linked to the observed lack of ROCK (Fig. 4.19A). This kinase plays an important role in the acquisition of the invasive phenotype of ovarian cancer cells (Han et al., 2005) and has therefore to be attributed to mesenchymal features. Interestingly, ROCK mRNA levels between R/E and S/M cultures were not significantly changed (Table 4.1), indicating a regulation of ROCK on the posttranscriptional level. A reported candidate for such regulation is MAPK or Erk kinase (MEK) (Pawlak and Helfman, 2002). Clearly, further studies would be needed, to delineate the impact of ROCK on the mesenchymal phenotype of ovc316 cultures and subsequently, on their resistance to adenovirus infection and oncolysis. However, attempts to introduce ROCK or constitutively active forms of Rho GTPases in R/E cells resulted either in poor transfection rates (<2%) or in massive cell death (>85%) after nucleofection, which made these approaches unsuitable (data not shown). Nevertheless, the inhibition of ROCK in S/M cultures could reduce transduction rates and dramatically increased survival of cells after treatment with Ad5/35.IR-E1A/TRAIL. The higher resistance to adenoviruses did not correlate with levels of E-cadherin, but was accompanied by an acquisition of the more epithelial, cobblestone-like morphology. Several studies have demonstrated that activation of RhoA in context with its effector kinase ROCK causes the disassembly of adherens junctions (Sahai and Marshall, 2002; Wojciak-Stothard et al., 2001). Therefore, one could hypothesize that cell-cell adhesion and tissue integrity in Rho and ROCK inhibitor treated S/M cells is more solid, despite the lack of E-cadherin and absence of a completely polarized R/E phenotype. This could, however, not fully inhibit the infection with adenovirus, but might have affected the release of progeny virus after infection with Ad5/35.IR-E1A/TRAIL. It is also possible that viral replication or virus assembly is

# DISCUSSION

impaired due to downstream effects of RhoA and/or ROCK in these cells. This hypothesis would be in line with the observation that viral oncolysis in S/M cells could also be significantly decreased by inhibition of Rho GTPases, including RhoA.

Considering the observed differences between ovc316-derived *in vitro* cultures and tumor xenografts, R/E cells seem to best resemble the *in vivo* phenotype of ovarian cancer. Despite different levels of ROCK, both cell populations were restricted to an epithelial phenotype and subsequently, adenoviral replication as well as oncolysis was inhibited. Furthermore, similar discrepancies for *in vitro* and *in vivo* cultures were also observed for cell lines derived from other epithelial cancers. As most tumor targeted adenoviral vectors have been selected and evaluated on cultured tumor cells, this raises the question of whether these viruses target the right phenotype and subsequently, whether they could actually infect and lyse the majority of cells in a given tumor. The finding that cells in tumor xenografts are almost exclusively in an epithelial or E/M hybrid stage could be even more dramatic when seen in context with a recent report, which shows that the loss of E-cadherin is induced by low levels of Rb (Arima et al., 2008). In this study the loss of Rb activated EMT inducers Slug and ZEB1, whereas its overexpression could inhibit EMT in human mammary epithelial cells. In epithelial restricted cells, this novel EMT-regulating role for the tumor-suppressor Rb could therefore abolish the oncolytic potential of E1-deleted adenoviruses that target Rb-deficient cells for tumor-specific replication. This would include all oncolytic viruses used throughout this thesis and could explain why basal infection of R/E cells with Ad5/35.IR-E1A/TRAIL did not result in efficient oncolysis (Fig. 4.19C), despite high transduction rates with GFP expressing viruses (Fig. 4.12E). Most likely due to the lack of EMT inducing signals, in ovc316 xenografts, a mesenchymal phenotype was almost absent (Fig. 4.14F). Subsequently, tumor transduction as well as adenoviral replication and subsequent oncolysis were almost completely inhibited. Adenovirus uptake of cells in tumor xenografts might be additionally compromised by degradation of FAK (Fig. 4.19A). Downstream signaling of FAK after adenovirus binding to integrins is an important initial step in the viral internalization process (Li et al., 1998b) and ovc316 xenografts mainly contained proteolytically modified versions of this kinase, which are known to be infunctional (Cooray et al., 1996). Also noteworthy is the fact, that multiple ovarian cancer markers including claudins 3, 4, and 7, E-cadherin, and EpCAM were associated with the resistant phenotype. All of these proteins are reported to increase with early progression of the disease and correlate with poor prognosis (Auersperg et al., 2002; Heinzelmann-Schwarz et al., 2004; Rangel et al., 2003; Tassi et al., 2008). The expression of epithelial junction markers E-cadherin or claudin 7 could also be confirmed for the majority of cells in patient biopsies or in xenografts derived from the ovarian cancer cell line SKOV3-ip1. The epithelial phenotype of ovarian cancer, which correlated with the expression of adenovirus receptors CD46, CAR, and $a_V$-integrins, can therefore be seen as a major barrier for commonly used oncolytic adenoviruses.

# DISCUSSION

## 5.3. Attempts to overcome resistance to adenoviral infection and oncolysis

### 5.3.1. Modification of the epithelial phenotype

As mentioned in 1.3.2., multiple pathways can individually or synergistically induce a mesenchymal phenotype in epithelial-derived cancer cells. Notably, EMT could be efficiently induced in R/E-EMT cultures during normal cell passaging. Concomitantly with the loss of the epithelial phenotype, these cultures lost the resistance to viral oncolysis (Fig. 4.7 and 4.19C). However, an apparent inability to undergo EMT under the same culturing conditions rendered R/E cultures resistant to viral oncolysis. In attempts to overcome this observed resistance, R/E cells were objected to culturing in the additional presence of several known EMT-inducing factors. These included treatment with growth factors FGF (fibroblast growth factor), TGF-$\beta$1 (transforming growth factor $\beta$1), IGF-1 (insulin-like growth factor 1), HGF (hepatocyte growth factor), and culturing on extracellular matrix components laminin, fibronectin, collagen I or IV (data not shown). Neither of these attempts succeeded in the induction of EMT nor in increase of adenovirus transduction. This implies that R/E cells were either refractory to signaling through the involved pathways or unable to execute morphological changes in response to them. It is therefore possible that R/E cells have mutations that intrinsically inhibit EMT, even if it is induced.

Nevertheless, a partial disruption of intercellular junctions in R/E cultures could be achieved by treatment with the bacterial toxin ZOT, which targets the underlying cytoskeleton. The F-actin network is the steric backbone of intercellular epithelial junctions and its alteration introduced plaques in R/E cell monolayers (Fig. 4.20A/B). However, GFP transduction as well as oncolytic adenovirus response was restricted to individual clusters within R/E cell layers, which can largely be attributed to lowered tissue integrity in these distinct areas due to ZOT treatment. In contrast, the majority of cells maintained intercellular junctions and were resistant to adenovirus-mediated transduction and oncolysis. A general breakdown of intercellular junctions could be achieved by extracellular depletion of calcium, which affected tissue integrity and subsequently lowered the number of surviving cells after infection with Ad5/35.IR-E1A/TRAIL (Fig. 4.21). Clearly, the decreased number of cells that initially remained attached to tissue culture plastic in versene treated wells influenced the outcome of this experiment. However, large clusters of versene-pretreated R/E cells remained alive in the presence of oncolytic adenovirus, implying a reestablishment of tissue integrity between these cells. Most likely, Ad5/35.IR-E1A/TRAIL predominantly killed cells that were reapplied prior to infection and inhibited the re-growth of the culture. Whereas calcium depletion and ZOT treatment led to morphological changes and could significantly enhance oncolytic virus performances, these approaches failed in the complete elimination of R/E cultures. It is therefore very likely that different degrees of resistance in response to tight junction modification and oncolytic virus treatment are present within cultures that are restricted to an epithelial phenotype. To overcome this resistance, EMT-inducing or tight junction-opening agents would have to be applied repetitively or in higher doses prior to adenovirus infection. In the case of ZOT, this was accompanied by undesired toxicity in the present study. Nevertheless, the described experiments show that a successful induction of EMT in initially epithelial-restricted cells can subsequently render affected cells susceptible to adenoviral infection and oncolysis.

### 5.3.2. Oncolytic potential of different adenovirus serotypes on epithelial restricted cells

Interestingly, several human viruses engage tight junction or other cell junction molecules to achieve entry into epithelial cells. Among these viruses is hepatitis C virus (Evans et al., 2007), reovirus (Barton et al., 2001), and herpes simplex virus (Geraghty et al., 1998). Of particular interest is coxsackievirus B, whose entry requires initial binding to a primary receptor (decay-accelerating factor) on the luminal cell surface, followed by lateral migration of the virus–receptor complex to the tight junction, where interaction with the CAR co-receptor and uptake into the host cell occurs (Coyne and Bergelson, 2006). Coxsackievirus B entry is accompanied by the specific internalization of occludin. As mentioned in 5.1, adenovirus serotype 5, which also targets CAR for attachment, is relatively inefficient in infection of airway epithelium because the virus is unable to attach to CAR when tight junctions are intact (Coyne and Bergelson, 2005). Experiments in the present thesis demonstrate that another major adenovirus receptor, CD46, is also trapped in tight junctions of epithelial cells, which limits the use of CD46-interacting adenoviruses for tumor gene therapy. While CAR- and CD46-targeted adenovirus serotypes use an infection mechanism that is relatively inefficient for entry into epithelial cells, another group of Ads (Ad3, 7, and 14) has evolved a different mechanism for cell infection, which is similar to that of coxsackie B virus (Tuve et al., 2008). It was recently showed, using Ad3 as an example, that this group of adenoviruses binds with low affinity to HSPGs. This binding is required to trigger high affinity Ad3 virus interaction to another cellular receptor, whereby the second interaction involves virus capsid proteins other than the fiber knob. This receptor has not been identified yet and therefore is tentatively named receptor X (Tuve et al., 2006). Notably, the HSPG/receptor X infection mechanism is also used by Ad11 when CD46 is absent or blocked like in the case of epithelial ovarian cancer cells. A number of parasites, for example Mycobacterium tuberculosis, also use HSPGs to enter cells without altering the integrity of tight junctions (Menozzi et al., 2006). The specific infection strategy of Ad3, 7, 11, and 14 might potentially explain why they are able to more efficiently lyse epithelial cancer cells than CAR- and CD46-interacting adenoviruses (Fig. 4.22). Most certainly, these adenovirus serotypes induced a 'knock on' effect in R/E cells by counteracting intracellular processes that have an inhibitory effect on EMT-inducing signaling and subsequent E-cadherin internalization. Considering a possible presence of Rb in R/E cultures (as mentioned in 5.2), wildtype adenoviruses could potentially counteract this activity by expression of E1A, which is known to be an inducer of EMT in human lung epithelial cells (Behzad et al., 2006). Another possible mechanism for the enhanced lytic ability of receptor X-targeted adenoviruses might be an excessive release of dodecahedra from initially infected R/E cells. The dodecahedron is a virus-like particle that is composed of either one (penton base) or two (penton base plus fiber) viral proteins of human adenovirus. Dodecahera are self-assembling, genome-less, and about four times smaller than fiberless viral capsids (Fender et al., 1997; Zochowska et al., 2009). Interestingly, the assembling of these particles is restricted to only a few adenovirus serotypes that do not include serotype 5 (Fender et al., 2003). The best-studied dodecahedron is derived from Ad3, where the ratio between infectious Ad3 particles and dodecahedra is $1:5.5 \times 10^6$ in infected cells (Fender et al., 2005). A recent study suggested that Ad3 dodecahedra bind to HSPGs which triggers subsequent interaction and internalization by cellular integrins (Fender et al., 2008). With respect to the superior oncolytic ability of adenoviruses targeted to receptor X, it is therefore possible that adenoviruses based on serotype 3, 7, 11 and 14 might use similar cell entry mechanisms, which may additionally employ dodecahedra. An excessive release of these virus-like particles after initial infection could therefore trigger morphological changes that are reminiscent of EMT in adjacent cells, which would subsequently render these susceptible to infection with adenovirus.

As an example for receptor X-targeted adenoviruses, Ad3 was tested for its oncolytic performance in ovc316 xenografts. In contrast to Ad35, Ad3 was able to replicate in subcutaneous tumors and significantly delayed tumor growth in immunodeficient mice (Fig. 4.21). Despite these promising results, it remains elusive how oncolytic vectors based on serotype Ad3, 7, 11 or 14 would perform in an immunocompetent model. That the immune system might be a serious concern indicates a study, where the host response against intratumorally injected human adenoviruses was evaluated in a syngeneic, immunocompetent breast cancer mouse model (Tuve et al., 2009). Injection of adenoviral vectors showed a strong T-cell mediated immune response against the vectors at tumor-draining lymph nodes and in the tumor microenvironment (tumor infiltrating lymphocytes). More importantly, all tested oncolytic and transgene-expressing adenoviruses showed anti-tumor efficacy similar to that of a replication-deficient and transgene-devoid adenoviral vector. This strongly suggests an anti-adenovirus immune response rather than oncolysis as a reason for tumor regression and highlights the immunogenic potential of adenoviruses in general. The study also questions whether oncolytic adenoviruses can efficiently replicate and release progeny virus before virus-harboring tumor cells are eliminated by the host's immune system.

## 5.4. Other anatomical and physical barriers within the tumor microenvironment

The finding that the epithelial phenotype of tumor cells is a limiting factor for infection and oncolysis is not the only barrier that has been reported for tumor targeted adenoviruses. As mentioned in 1.4.4., several blood components, as well as the ECM secreted by tumor stroma cells dramatically lower the effective viral dose before oncolytic adenoviruses reach their target cells. Interestingly, the present study shows that tumor cells with mesenchymal features can also produce excessive amounts of ECM compounds, such as laminin, fibronectin, and collagen IV. In the following paragraphs further anatomical and physical barriers that could potentially hinder adenoviral spread in solid tumors will be discussed.

### 5.4.1. Transendothelial transport

Only limited knowledge is available about viral spread in solid tumors, but much can be learned from models that were generated to study the distribution of macromolecular drugs. Once systemically administered viruses reach distant tumor sites, they have to exit the tumor vasculature and travel through the interstitial space in order to reach their target cells. The endothelial cell layer, which lines the blood vessels, is thought to represent a barrier to adenoviral vectors (Fechner et al., 1999). To date, its impact on viral distribution in tumors remains relatively unclear. The size of adenoviruses (about 90 nm) suggests that they would benefit from a phenomenon known as the enhanced permeability and retention (EPR) effect in solid tumors. The EPR effect is observed for intravenously administered macromolecular anticancer drugs that escape renal clearance, due to their large molecular size (10-500 nm). They are mostly unable to pass the tight endothelial junctions of normal blood vessels, but can extravasate and then become trapped in the tumor vicinity (Maeda et al., 2000). Unlike normal tissues that feature an organized vascular network, the blood vessel system in solid tumors is rather chaotic. The endothelial cell layers are poorly aligned (Iyer et al., 2006) and elevated levels of vascular permeability factors generate "leaky" capillaries (Maeda et al., 2000). Based on this, one could estimate that transendothelial transport of adenoviral particles is not a limiting obstacle for tumor cell transduction. In this context, it is notable that the endothelial cell layer in hepatic capillaries, in contrast to many other tissues and organs, is fenestrated. Therefore, this anatomical feature, in part, can also explain the preferential transduction of liver cells after intravenous adenovirus injection (Wisse et al., 2008).

### 5.4.2. Intratumoral pressure

Rapid tumor cell proliferation and weakly developed lymphatics cause high interstitial fluid pressure (IFP) (Jain, 1997; Milosevic et al., 1998) and blood vessel remodeling by intussusception (Patan et al., 1996) or compression (Padera et al., 2004). Additionally, the increased hydraulic conductivity of "leaky" capillaries can further increase the IFP in tumors (Jain et al., 2007). Together, this leads to an imbalance in blood flow and nutrient supply within the tumor microenvironment. The uniformly high IFP in the center of solid tumors drops towards the periphery (Boucher et al., 1990), which could additionally lower fluid extravasations and the delivery of viral particles in the high-pressure regions. Moreover, the vascular surface area per unit tissue weight is decreasing with tumor growth, which further limits the transvascular exchange for large tumors compared to small tumors (Baxter and Jain, 1989; Baxter and Jain, 1990). Cells that are distant to blood vessels (100-200 mm) and located in high pressure regions subsequently constitute large areas of hypoxic, necrotic or semi-necrotic tissue. This exacerbates the tendency of tumor cells to overproduce and release lactic acids within these regions, which results in low extracellular pH values (acidosis) (Gatenby and Gillies, 2004). In contrast, cells situated in the proximity of stabilized microfluid circulation and in the invasive front benefit from the enhanced vascular permeability that supplies adequate amounts of macromolecules for rapid tumor growth (Jain, 1998). Furthermore, the blood flow rates in non-necrotic regions can be substantially higher than in the surrounding normal tissue (Sevick and Jain, 1991). It is therefore expectable that the uptake of adenoviruses in solid tumors is heterogeneous and the general distribution might decrease with increasing tumor weight.

Supporting this theory, a number of reports show that intravenously or intraperitonially delivered adenoviruses can be primarily found in areas next to tumor blood vessels or neighboring tumor stroma while the anti-tumor efficacy decreases with gain in tumor weight (Shayakhmetov et al., 2002; Shinozaki et al., 2006).

Several studies in animals and humans have demonstrated a strong synergistic benefit for oncolytic adenovirus treatments when combined with chemotherapy or radiation after intratumoral injections (Crompton and Kirn, 2007; Raki et al., 2005; Rogulski et al., 2000; Yu and Fang, 2007). This phenomenon could have multiple reasons. First, adenoviruses can replicate in and efficiently lyse proliferating and non-proliferating cells, whereas most chemotherapeutic agents and radiation therapy target cycling cells (Fukumura and Jain, 2007). Second, the induction of apoptosis can enhance the viral spread in tumors (Mi et al., 2001; Nagano et al., 2008) and third, the induced massive cell death can lower the IFP in tumors, which was shown to enhance spread and efficacy of oncolytic herpes simplex virus in tumors (Taghian et al., 2005). In this context, the repetitive injection of oncolytic adenoviruses into solid tumors might also be beneficial for viral spread due to lowered IFP after initial cell death. The application of chemotherapy to lower the IFP is used in approaches to 'normalize' the tumor vasculature. Anti-angiogenic drugs are thought to compensate the pro-angiogenic factors that are extensively produced in the tumor in order to eliminate "leaky" blood vessels. Ideally, this would lead to a more organized blood vessel system that features more functional and more uniformly perfused capillaries within solid tumors. The 'normalization' approach might offer a timely window for better distribution of therapeutic agents and viruses, but to date remains controversial (Jain, 2008).

### 5.4.3. Acidosis and hypoxia

Two major consequences of abnormal microcirculation in solid tumors are hypoxia and low extracellular pH. Both metabolic processes induce the expression of angiogenic factors in a complimentary manner and can trigger tumor cells to be metastatic and invasive (Fukumura and Jain, 2007). How the low pH environment of solid tumors affects the adenoviral infection and replication process is largely unknown, but several *in vitro* observations indicate significant alterations. An early study done on HeLa cells suggested an aberrant, fiber independent attachment process for Ad2 particles in acidic medium (Svensson, 1985). However, internalization of viral capsids was still efficient. Importantly, the acidification of endosomes is a critical step in the adenovirus infection process that facilitates the escape from internalized viral capsids by endocytic vesicle lysis (Seth et al., 1984a). During investigations analyzing this mechanism, it was also found that adenoviruses can simply disrupt the plasma membranes of cells, if these are maintained in acidic pH (Seth et al., 1984b). Lately, this direct lytic activity was linked to a conformational change of viral capsids at pH values below 5.5 (Wiethoff et al., 2005). Prior to membrane penetration, peripentonal hexons are removed from the adenovirus capsid. This subsequently exposes the originally hidden protein VI that confers the membrane lysing function. While these findings suggest efficient viral uptake into cancer cells in an acidic environment, they also indicate the loss of the fiber protein in response to acidification, which would ultimately affect receptor targeting. Potential approaches to avoid virus capsid-degradation, i.e. the loss of fiber, in the acidic microenvironment and to prolong the intratumoral persistence of adenoviruses include virus particle coating with synthetic polymers or polyethylene glycol (Hatefi et al., 2007; Singh et al., 2008). While such approaches increase the persistence of viral vectors in the tumor microenvironment and might offer reduced immunogenicity, they can also impede the natural, receptor-targeted infection and internalization mechanism of adenoviruses.

Hypoxia affects adenoviral replication, due to lowered production of E1A and fiber proteins (Pipiya et al., 2005). In infected hypoxic tumor cells, this would drastically lower the release of progeny virus and consequently vectors addressing this problem have been developed (Cho et al., 2004; Post et al., 2004). A recent study shows that an oncolytic adenovirus that replicates in a hypoxia-dependent manner and also expresses IL-12 can enhance anti-tumor efficacy and transgene expression when compared to non-replicating vectors in an immunocompetent Syrian hamster model (Bortolanza et al., 2009).

## 5.5. Conclusions and future directions

The present thesis sheds light on an obstacle for oncolytic adenoviruses that had not been taken into account so far. During the course of this study it was identified that the targeted cancer cell itself does not support adenoviral infection and oncolysis when it is restricted to an epithelial phenotype (see figure 5.1 for a detailed description). Most likely in synergy with other limiting factors, the epithelial phenotype of cancer cells affects adenoviral spread, infection, and replication in solid tumors. Particularly, two major physical barriers, namely tumor ECM and intercellular tight junctions, appear to form an unfortunate barrier to tumor targeted adenoviruses. Epithelial restricted cells and cells with mesenchymal features seem to complement each other, since the latter are able to produce excessive amounts of ECM while in the process of losing the tight junction barrier. In solid tumors, this could additionally shield epithelial tumor cell nests that are still sealed by their intercellular junctions. An important conclusion from this thesis is that mesenchymal tumors (e.g. sarcomas) might be a better target for current oncolytic adenoviruses. Furthermore, the link of intracellular mesenchymal features to adenoviral susceptibility gives a rationale for the controlled induction of EMT in solid tumors prior to viral administration. This might be a double-edged sword though, because EMT enforces metastatic potential and recently has been linked to cancer stem cells (Polyak and Weinberg, 2009). At the same time, it is intriguing that specific adenovirus serotypes seem to target exactly these cells in solid tumors, which are thought to be responsible for metastatic spread and tumor re-growth after common treatments. Theoretically, this could make oncolytic adenoviruses a powerful weapon in the arsenal of therapeutics against epithelial cancers. Given their superior performance on epithelial tissue, future oncolytic vectors should be based on serotypes derived from subgroup B adenoviruses that target receptors other than CAR or CD46. Here, it would also be interesting to know, whether all adenoviruses that target receptor X assemble and release dodecahedra, as it is described for Ad3. Furthermore, it should be delineated, whether the release of dodecahedra or other factors are responsible for the induction of EMT-like processes in epithelial-restricted cells after viral infection with these adenovirus serotypes. In this context, it might also be beneficial to combine the use of oncolytic viruses based on e.g. Ad3 or Ad11 with chemotherapy. The induction of EMT-like processes in solid tumors due to the viral infection process and/or release of dodecahedra from infected cells could induce a bystander-effect in solid tumors that might increase drug absorption through epithelial layers. Furthermore, the observed differences between *in vitro* and *in vivo* phenotypes of several epithelial cancer cell lines due to EMT, strongly suggest a re-evaluation of current strategies in targeting of adenoviral infection and replication to tumor cells. As discussed, specifically, the targeting to cancer cells with defective Rb pathways might be impaired to a large extend in epithelial tumors. With respect to this, the expression of E1A under promoters that are known to be active in solid tumors could be beneficial. However, a systematic analysis of Rb levels in solid tumors or tumor xenografts seams necessary in order to draw further conclusions.

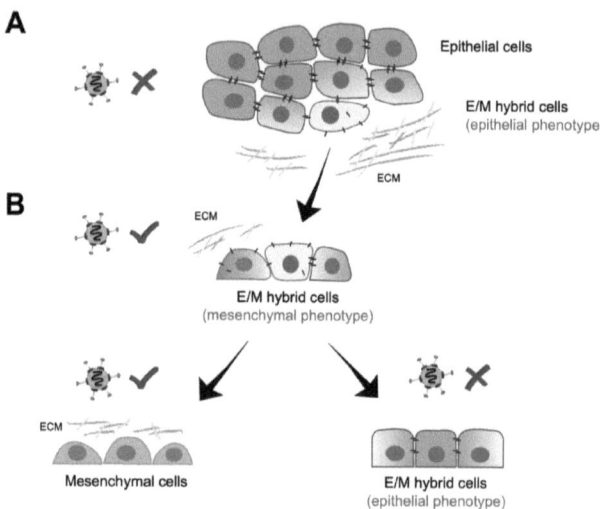

**Figure 5.1: Phenotypes of ovarian cancer cells and their susceptibility to oncolytic adenoviruses. A)** Tumors or tumor xenografts contain almost exclusively epithelial and epithelial/mesenchymal (E/M) hybrid cells, which are restricted to an epithelial phenotype. These cells are resistant to adenovirus infection and oncolysis. **B)** E/M hybrid cells that adapt to tissue culture mostly have a mesenchymal phenotype and generate mesenchymal (S/M) cells during passaging. These *in vitro* cultures comprise cells susceptible to adenovirus infection and oncolysis. A minority of epithelial-restricted (R/E) cells among E/M hybrid cells *in vitro* can be isolated and propagated by clonal cell expansion. ECM=extracellular matrix, √=susceptible, X=resistant

Clearly, new culturing models that better reflect the situation of cancer cells in tumors are urgently needed to further delineate their interaction with adenoviruses. These models should ideally employ a three-dimensional complexity, which includes tumor stroma and additionally, mirror the metabolic conditions present within solid tumors. In this context, it will be of special importance to identify the specific underlying mechanisms of why R/E cells are not able to undergo EMT. This knowledge could enable a controlled maintenance of the epithelial phenotype *in vitro* and thereby avoid an undesired EMT in primary cultures. This thesis also contains several findings that are interesting in context with the general biology of cancer. Specifically, the fact that only E/M cells, but not epithelial cells within ovarian cancer xenografts can adapt to tissue culture conditions leads to a discrepancy between phenotypes of cancer cells *in vivo* and *in vitro*, as noted above. This raises the question, how well differentiated epithelial cells can be maintained *in vitro* and subsequently, whether all epithelial cells can actually undergo an EMT. Due to a lack of pathohistological evidence, EMT was long only recognized as a tissue culture phenomenon. It might therefore already require a number of initial mesenchymal traits (e.g. vimentin) in epithelial cells in order to enable a transdifferentiation into the mesenchymal lineage. In a different study using ovc316 cells, it could also be showed that tumor formation is depending on the presence of E/M cells, which can generate epithelial cells *in vivo* and mesenchymal cells *in vitro*. On the contrary, clonal R/E and S/M cultures, which never switched lineages, had no tumor forming abilities in immunocompromised mice (Strauss et al., manuscript in preparation).

## 6. References

Adams, C. L., Chen, Y. T., Smith, S. J., and Nelson, W. J. (1998). Mechanisms of epithelial cell-cell adhesion and cell compaction revealed by high-resolution tracking of E-cadherin-green fluorescent protein. J Cell Biol *142*, 1105-1119.

Agarwal, R., D'Souza, T., and Morin, P. J. (2005). Claudin-3 and claudin-4 expression in ovarian epithelial cells enhances invasion and is associated with increased matrix metalloproteinase-2 activity. Cancer Res *65*, 7378-7385.

Ahmed, N., Maines-Bandiera, S., Quinn, M. A., Unger, W. G., Dedhar, S., and Auersperg, N. (2006). Molecular pathways regulating EGF-induced epithelio-mesenchymal transition in human ovarian surface epithelium. Am J Physiol Cell Physiol *290*, C1532-1542.

Aho, S., Levansuo, L., Montonen, O., Kari, C., Rodeck, U., and Uitto, J. (2002). Specific sequences in p120ctn determine subcellular distribution of its multiple isoforms involved in cellular adhesion of normal and malignant epithelial cells. J Cell Sci *115*, 1391-1402.

Alemany, R. (2007). Cancer selective adenoviruses. Mol Aspects Med *28*, 42-58.

Anders, M., Hansen, R., Ding, R. X., Rauen, K. A., Bissell, M. J., and Korn, W. M. (2003). Disruption of 3D tissue integrity facilitates adenovirus infection by deregulating the coxsackievirus and adenovirus receptor. Proc Natl Acad Sci U S A *100*, 1943-1948.

Andreeva, A. Y., Krause, E., Muller, E. C., Blasig, I. E., and Utepbergenov, D. I. (2001). Protein kinase C regulates the phosphorylation and cellular localization of occludin. J Biol Chem *276*, 38480-38486.

Aparicio, O., Razquin, N., Zaratiegui, M., Narvaiza, I., and Fortes, P. (2006). Adenovirus virus-associated RNA is processed to functional interfering RNAs involved in virus production. J Virol *80*, 1376-1384.

Arima, Y., Inoue, Y., Shibata, T., Hayashi, H., Nagano, O., Saya, H., and Taya, Y. (2008). Rb depletion results in deregulation of E-cadherin and induction of cellular phenotypic changes that are characteristic of the epithelial-to-mesenchymal transition. Cancer Res *68*, 5104-5112.

Arnoux, V., Come, C., Kusewitt, D., Hudson, L., and Savagner, P. (2005). Cutaneous Wound Reepithelization: A partial and reversible EMT. Springer, Berlin *Rise and Fall of epithelial phenotype: concepts of epithelial-mesenchymal transition.*, 111-134.

Asada, T. (1974). Treatment of human cancer with mumps virus. Cancer *34*, 1907-1928.

Auersperg, N., Maines-Bandiera, S. L., Dyck, H. G., and Kruk, P. A. (1994). Characterization of cultured human ovarian surface epithelial cells: phenotypic plasticity and premalignant changes. Lab Invest *71*, 510-518.

Auersperg, N., Ota, T., and Mitchell, G. W. (2002). Early events in ovarian epithelial carcinogenesis: progress and problems in experimental approaches. Int J Gynecol Cancer *12*, 691-703.

Auersperg, N., Woo, M. M., and Gilks, C. B. (2008). The origin of ovarian carcinomas: a developmental view. Gynecol Oncol *110*, 452-454.

Aunoble, B., Sanches, R., Didier, E., and Bignon, Y. J. (2000). Major oncogenes and tumor suppressor genes involved in epithelial ovarian cancer (review). Int J Oncol *16*, 567-576.

Barker, D. D., and Berk, A. J. (1987). Adenovirus proteins from both E1B reading frames are required for transformation of rodent cells by viral infection and DNA transfection. Virology *156*, 107-121.

# REFERENCES

Barry, F. P., and Murphy, J. M. (2004). Mesenchymal stem cells: clinical applications and biological characterization. The international journal of biochemistry & cell biology 36, 568-584.

Barton, E. S., Forrest, J. C., Connolly, J. L., Chappell, J. D., Liu, Y., Schnell, F. J., Nusrat, A., Parkos, C. A., and Dermody, T. S. (2001). Junction adhesion molecule is a receptor for reovirus. Cell 104, 441-451.

Batlle, E., Sancho, E., Franci, C., Dominguez, D., Monfar, M., Baulida, J., and Garcia De Herreros, A. (2000). The transcription factor snail is a repressor of E-cadherin gene expression in epithelial tumour cells. Nat Cell Biol 2, 84-89.

Baxter, L. T., and Jain, R. K. (1989). Transport of fluid and macromolecules in tumors. I. Role of interstitial pressure and convection. Microvasc Res 37, 77-104.

Baxter, L. T., and Jain, R. K. (1990). Transport of fluid and macromolecules in tumors. II. Role of heterogeneous perfusion and lymphatics. Microvasc Res 40, 246-263.

Behzad, A. R., Morimoto, K., Gosselink, J., Green, J., Hogg, J. C., and Hayashi, S. (2006). Induction of mesenchymal cell phenotypes in lung epithelial cells by adenovirus E1A. Eur Respir J 28, 1106-1116.

Belin, M. T., and Boulanger, P. (1987). Processing of vimentin occurs during the early stages of adenovirus infection. J Virol 61, 2559-2566.

Bell, D. A. (2005). Origins and molecular pathology of ovarian cancer. Mod Pathol 18 Suppl 2, S19-32.

Berchuck, A., Iversen, E. S., Lancaster, J. M., Pittman, J., Luo, J., Lee, P., Murphy, S., Dressman, H. K., Febbo, P. G., West, M., et al. (2005). Patterns of gene expression that characterize long-term survival in advanced stage serous ovarian cancers. Clin Cancer Res 11, 3686-3696.

Bergelson, J. M., Cunningham, J. A., Droguett, G., Kurt-Jones, E. A., Krithivas, A., Hong, J. S., Horwitz, M. S., Crowell, R. L., and Finberg, R. W. (1997). Isolation of a common receptor for Coxsackie B viruses and adenoviruses 2 and 5. Science 275, 1320-1323.

Berk, A. J. (2007). Adenoviridae: the viruses and their replication. Fields Virology, 5th edition In D. M. Knipe, P. M. Howley, D. E. Griffin, R. A. Lamb, M. A. Martin, B. Roizman, and S. E. Straus (ed.), 2355-2394

Berkner, K. L., and Sharp, P. A. (1984). Expression of dihydrofolate reductase, and of the adjacent Elb region, in an Ad5-dihydrofolate reductase recombinant virus. Nucleic Acids Res 12, 1925-1941.

Bernt, K., Liang, M., Ye, X., Ni, S., Li, Z. Y., Ye, S. L., Hu, F., and Lieber, A. (2002). A new type of adenovirus vector that utilizes homologous recombination to achieve tumor-specific replication. J Virol 76, 10994-11002.

Berx, G., Cleton-Jansen, A. M., Strumane, K., de Leeuw, W. J., Nollet, F., van Roy, F., and Cornelisse, C. (1996). E-cadherin is inactivated in a majority of invasive human lobular breast cancers by truncation mutations throughout its extracellular domain. Oncogene 13, 1919-1925.

Berx, G., and Van Roy, F. (2001). The E-cadherin/catenin complex: an important gatekeeper in breast cancer tumorigenesis and malignant progression. Breast Cancer Res 3, 289-293.

Bett, A. J., Haddara, W., Prevec, L., and Graham, F. L. (1994). An efficient and flexible system for construction of adenovirus vectors with insertions or deletions in early regions 1 and 3. Proc Natl Acad Sci U S A 91, 8802-8806.

Bett, A. J., Krougliak, V., and Graham, F. L. (1995). DNA sequence of the deletion/insertion in early region 3 of Ad5 dl309. Virus Res 39, 75-82.

# REFERENCES

Birchmeier, W., and Behrens, J. (1994). Cadherin expression in carcinomas: role in the formation of cell junctions and the prevention of invasiveness. Biochim Biophys Acta *1198*, 11-26.

Bischoff, J. R., Kirn, D. H., Williams, A., Heise, C., Horn, S., Muna, M., Ng, L., Nye, J. A., Sampson-Johannes, A., Fattaey, A., and McCormick, F. (1996). An adenovirus mutant that replicates selectively in p53-deficient human tumor cells. Science *274*, 373-376.

Bjorge, L., Junnikkala, S., Kristoffersen, E. K., Hakulinen, J., Matre, R., and Meri, S. (1997). Resistance of ovarian teratocarcinoma cell spheroids to complement-mediated lysis. Br J Cancer *75*, 1247-1255.

Bortolanza, S., Bunuales, M., Otano, I., Gonzalez-Aseguinolaza, G., Ortiz-de-Solorzano, C., Perez, D., Prieto, J., and Hernandez-Alcoceba, R. (2009). Treatment of pancreatic cancer with an oncolytic adenovirus expressing interleukin-12 in Syrian hamsters. Mol Ther *17*, 614-622.

Boucher, Y., Baxter, L. T., and Jain, R. K. (1990). Interstitial pressure gradients in tissue-isolated and subcutaneous tumors: implications for therapy. Cancer Res *50*, 4478-4484.

Brabletz, T., Jung, A., Reu, S., Porzner, M., Hlubek, F., Kunz-Schughart, L. A., Knuechel, R., and Kirchner, T. (2001). Variable beta-catenin expression in colorectal cancers indicates tumor progression driven by the tumor environment. Proceedings of the National Academy of Sciences of the United States of America *98*, 10356-10361.

Brabletz, T., Jung, A., Spaderna, S., Hlubek, F., and Kirchner, T. (2005). Opinion: migrating cancer stem cells - an integrated concept of malignant tumour progression. Nat Rev Cancer *5*, 744-749.

Braga, V. (2000). Epithelial cell shape: cadherins and small GTPases. Exp Cell Res *261*, 83-90.

Brand, K., Loser, P., Arnold, W., Bartels, T., and Strauss, M. (1998). Tumor cell-specific transgene expression prevents liver toxicity of the adeno-HSVtk/GCV approach. Gene Ther *5*, 1363-1371.

Brown, E., McKee, T., diTomaso, E., Pluen, A., Seed, B., Boucher, Y., and Jain, R. K. (2003). Dynamic imaging of collagen and its modulation in tumors in vivo using second-harmonic generation. Nat Med *9*, 796-800.

Bruder, J. T., Appiah, A., Kirkman, W. M., 3rd, Chen, P., Tian, J., Reddy, D., Brough, D. E., Lizonova, A., and Kovesdi, I. (2000). Improved production of adenovirus vectors expressing apoptotic transgenes. Hum Gene Ther *11*, 139-149.

Bryant, D. M., Wylie, F. G., and Stow, J. L. (2005). Regulation of endocytosis, nuclear translocation, and signaling of fibroblast growth factor receptor 1 by E-cadherin. Mol Biol Cell *16*, 14-23.

Caca, K., Kolligs, F. T., Ji, X., Hayes, M., Qian, J., Yahanda, A., Rimm, D. L., Costa, J., and Fearon, E. R. (1999). Beta- and gamma-catenin mutations, but not E-cadherin inactivation, underlie T-cell factor/lymphoid enhancer factor transcriptional deregulation in gastric and pancreatic cancer. Cell Growth Differ *10*, 369-376.

Cannito, S., Novo, E., Compagnone, A., Valfre di Bonzo, L., Busletta, C., Zamara, E., Paternostro, C., Povero, D., Bandino, A., Bozzo, F., *et al.* (2008). Redox mechanisms switch on hypoxia-dependent epithelial-mesenchymal transition in cancer cells. Carcinogenesis *29*, 2267-2278.

Cano, A., and Nieto, M. A. (2008). Non-coding RNAs take centre stage in epithelial-to-mesenchymal transition. Trends Cell Biol *18*, 357-359.

Capaldo, C. T., and Macara, I. G. (2007). Depletion of E-cadherin disrupts establishment but not maintenance of cell junctions in Madin-Darby canine kidney epithelial cells. Mol Biol Cell *18*, 189-200.

# REFERENCES

Chaffer, C. L., Thompson, E. W., and Williams, E. D. (2007). Mesenchymal to epithelial transition in development and disease. Cells Tissues Organs *185*, 7-19.

Chartier, C., Degryse, E., Gantzer, M., Dieterle, A., Pavirani, A., and Mehtali, M. (1996). Efficient generation of recombinant adenovirus vectors by homologous recombination in Escherichia coli. J Virol *70*, 4805-4810.

Chen, M., Chen, L. M., and Chai, K. X. (2006). Androgen regulation of prostasin gene expression is mediated by sterol-regulatory element-binding proteins and SLUG. Prostate *66*, 911-920.

Chen, P. H., Ornelles, D. A., and Shenk, T. (1993). The adenovirus L3 23-kilodalton proteinase cleaves the amino-terminal head domain from cytokeratin 18 and disrupts the cytokeratin network of HeLa cells. J Virol *67*, 3507-3514.

Chen, Y. T., Stewart, D. B., and Nelson, W. J. (1999). Coupling assembly of the E-cadherin/beta-catenin complex to efficient endoplasmic reticulum exit and basal-lateral membrane targeting of E-cadherin in polarized MDCK cells. J Cell Biol *144*, 687-699.

Cheng, J., Sauthoff, H., and Hay, J. G. (2007a). How do changes in tumor matrix affect the outcome of virotherapy? Cancer Biol Ther *6*, 290-292.

Cheng, J., Sauthoff, H., Huang, Y., Kutler, D. I., Bajwa, S., Rom, W. N., and Hay, J. G. (2007b). Human matrix metalloproteinase-8 gene delivery increases the oncolytic activity of a replicating adenovirus. Mol Ther *15*, 1982-1990.

Chiocca, E. A., Abbed, K. M., Tatter, S., Louis, D. N., Hochberg, F. H., Barker, F., Kracher, J., Grossman, S. A., Fisher, J. D., Carson, K., *et al.* (2004). A phase I open-label, dose-escalation, multi-institutional trial of injection with an E1B-Attenuated adenovirus, ONYX-015, into the peritumoral region of recurrent malignant gliomas, in the adjuvant setting. Mol Ther *10*, 958-966.

Cho, E. Y., Choi, Y., Chae, S. W., Sohn, J. H., and Ahn, G. H. (2006). Immunohistochemical study of the expression of adhesion molecules in ovarian serous neoplasms. Pathol Int *56*, 62-70.

Cho, W. K., Seong, Y. R., Lee, Y. H., Kim, M. J., Hwang, K. S., Yoo, J., Choi, S., Jung, C. R., and Im, D. S. (2004). Oncolytic effects of adenovirus mutant capable of replicating in hypoxic and normoxic regions of solid tumor. Mol Ther *10*, 938-949.

Christ, B., and Ordahl, C. P. (1995). Early stages of chick somite development. Anat Embryol (Berl) *191*, 381-396.

Christiansen, J. J., and Rajasekaran, A. K. (2006). Reassessing epithelial to mesenchymal transition as a prerequisite for carcinoma invasion and metastasis. Cancer Res *66*, 8319-8326.

Cohen, C. J., Shieh, J. T., Pickles, R. J., Okegawa, T., Hsieh, J. T., and Bergelson, J. M. (2001). The coxsackievirus and adenovirus receptor is a transmembrane component of the tight junction. Proc Natl Acad Sci U S A *98*, 15191-15196.

Come, C., Magnino, F., Bibeau, F., De Santa Barbara, P., Becker, K. F., Theillet, C., and Savagner, P. (2006). Snail and slug play distinct roles during breast carcinoma progression. Clin Cancer Res *12*, 5395-5402.

Comijn, J., Berx, G., Vermassen, P., Verschueren, K., van Grunsven, L., Bruyneel, E., Mareel, M., Huylebroeck, D., and van Roy, F. (2001). The two-handed E box binding zinc finger protein SIP1 downregulates E-cadherin and induces invasion. Mol Cell *7*, 1267-1278.

Conacci-Sorrell, M., Zhurinsky, J., and Ben-Ze'ev, A. (2002). The cadherin-catenin adhesion system in signaling and cancer. J Clin Invest *109*, 987-991.

# REFERENCES

Conget, P. A., and Minguell, J. J. (1999). Phenotypical and functional properties of human bone marrow mesenchymal progenitor cells. Journal of cellular physiology *181*, 67-73.

Cooray, P., Yuan, Y., Schoenwaelder, S. M., Mitchell, C. A., Salem, H. H., and Jackson, S. P. (1996). Focal adhesion kinase (pp125FAK) cleavage and regulation by calpain. Biochem J *318 ( Pt 1)*, 41-47.

Cox, D. S., Raje, S., Gao, H., Salama, N. N., and Eddington, N. D. (2002). Enhanced permeability of molecular weight markers and poorly bioavailable compounds across Caco-2 cell monolayers using the absorption enhancer, zonula occludens toxin. Pharmaceutical research *19*, 1680-1688.

Coyne, C. B., and Bergelson, J. M. (2005). CAR: a virus receptor within the tight junction. Adv Drug Deliv Rev *57*, 869-882.

Coyne, C. B., and Bergelson, J. M. (2006). Virus-induced Abl and Fyn kinase signals permit coxsackievirus entry through epithelial tight junctions. Cell *124*, 119-131.

Crijns, A. P., Gerbens, F., Plantinga, A. E., Meersma, G. J., de Jong, S., Hofstra, R. M., de Vries, E. G., van der Zee, A. G., de Bock, G. H., and te Meerman, G. J. (2006). A biological question and a balanced (orthogonal) design: the ingredients to efficiently analyze two-color microarrays with Confirmatory Factor Analysis. BMC Genomics *7*, 232.

Croft, D. R., Sahai, E., Mavria, G., Li, S., Tsai, J., Lee, W. M., Marshall, C. J., and Olson, M. F. (2004). Conditional ROCK activation in vivo induces tumor cell dissemination and angiogenesis. Cancer Res *64*, 8994-9001.

Crompton, A. M., and Kirn, D. H. (2007). From ONYX-015 to armed vaccinia viruses: the education and evolution of oncolytic virus development. Curr Cancer Drug Targets *7*, 133-139.

Crouzet, J., Naudin, L., Orsini, C., Vigne, E., Ferrero, L., Le Roux, A., Benoit, P., Latta, M., Torrent, C., Branellec, D., *et al.* (1997). Recombinational construction in Escherichia coli of infectious adenoviral genomes. Proc Natl Acad Sci U S A *94*, 1414-1419.

Crum, C. P., Drapkin, R., Kindelberger, D., Medeiros, F., Miron, A., and Lee, Y. (2007). Lessons from BRCA: the tubal fimbria emerges as an origin for pelvic serous cancer. Clin Med Res *5*, 35-44.

Darai, E., Bringuier, A. F., Walker-Combrouze, F., Feldmann, G., Madelenat, P., and Scoazec, J. Y. (1998). Soluble adhesion molecules in serum and cyst fluid from patients with cystic tumours of the ovary. Hum Reprod *13*, 2831-2835.

Davies, B. R., Worsley, S. D., and Ponder, B. A. (1998). Expression of E-cadherin, alpha-catenin and beta-catenin in normal ovarian surface epithelium and epithelial ovarian cancers. Histopathology *32*, 69-80.

Davis, M. A., Ireton, R. C., and Reynolds, A. B. (2003). A core function for p120-catenin in cadherin turnover. J Cell Biol *163*, 525-534.

Davison, E., Kirby, I., Whitehouse, J., Hart, I., Marshall, J. F., and Santis, G. (2001). Adenovirus type 5 uptake by lung adenocarcinoma cells in culture correlates with Ad5 fibre binding is mediated by alpha(v)beta1 integrin and can be modulated by changes in beta1 integrin function. J Gene Med *3*, 550-559.

de Jong, J. C., Osterhaus, A. D., Jones, M. S., and Harrach, B. (2008). Human adenovirus type 52: a type 41 in disguise? J Virol *82*, 3809; author reply 3809-3810.

DePage, N. G. (1912). Sulla scomparsa di un enorme canero vegetante del callo dell'utero senza cura chirurgica. Ginecologia, 82-88.

# REFERENCES

DeWeese, T. L., van der Poel, H., Li, S., Mikhak, B., Drew, R., Goemann, M., Hamper, U., DeJong, R., Detorie, N., Rodriguez, R., et al. (2001). A phase I trial of CV706, a replication-competent, PSA selective oncolytic adenovirus, for the treatment of locally recurrent prostate cancer following radiation therapy. Cancer Res 61, 7464-7472.

Doronin, K., Toth, K., Kuppuswamy, M., Ward, P., Tollefson, A. E., and Wold, W. S. (2000). Tumor-specific, replication-competent adenovirus vectors overexpressing the adenovirus death protein. J Virol 74, 6147-6155.

Drees, F., Pokutta, S., Yamada, S., Nelson, W. J., and Weis, W. I. (2005). Alpha-catenin is a molecular switch that binds E-cadherin-beta-catenin and regulates actin-filament assembly. Cell 123, 903-915.

Dressman, H. K., Berchuck, A., Chan, G., Zhai, J., Bild, A., Sayer, R., Cragun, J., Clarke, J., Whitaker, R. S., Li, L., et al. (2007). An integrated genomic-based approach to individualized treatment of patients with advanced-stage ovarian cancer. J Clin Oncol 25, 517-525.

du Bois, A., Quinn, M., Thigpen, T., Vermorken, J., Avall-Lundqvist, E., Bookman, M., Bowtell, D., Brady, M., Casado, A., Cervantes, A., et al. (2005). 2004 consensus statements on the management of ovarian cancer: final document of the 3rd International Gynecologic Cancer Intergroup Ovarian Cancer Consensus Conference (GCIG OCCC 2004). Ann Oncol 16 Suppl 8, viii7-viii12.

Dubeau, L. (1999). The cell of origin of ovarian epithelial tumors and the ovarian surface epithelium dogma: does the emperor have no clothes? Gynecol Oncol 72, 437-442.

Dumont, N., Wilson, M. B., Crawford, Y. G., Reynolds, P. A., Sigaroudinia, M., and Tlsty, T. D. (2008). Sustained induction of epithelial to mesenchymal transition activates DNA methylation of genes silenced in basal-like breast cancers. Proc Natl Acad Sci U S A 105, 14867-14872.

Duque, P. M., Alonso, C., Sanchez-Prieto, R., Lleonart, M., Martinez, C., de Buitrago, G. G., Cano, A., Quintanilla, M., and Ramon y Cajal, S. (1999). Adenovirus lacking the 19-kDa and 55-kDa E1B genes exerts a marked cytotoxic effect in human malignant cells. Cancer Gene Ther 6, 554-563.

Eger, A., Aigner, K., Sonderegger, S., Dampier, B., Oehler, S., Schreiber, M., Berx, G., Cano, A., Beug, H., and Foisner, R. (2005). DeltaEF1 is a transcriptional repressor of E-cadherin and regulates epithelial plasticity in breast cancer cells. Oncogene 24, 2375-2385.

Ellenberger, T., Fass, D., Arnaud, M., and Harrison, S. C. (1994). Crystal structure of transcription factor E47: E-box recognition by a basic region helix-loop-helix dimer. Genes Dev 8, 970-980.

Engelhardt, J. F., Ye, X., Doranz, B., and Wilson, J. M. (1994). Ablation of E2A in recombinant adenoviruses improves transgene persistence and decreases inflammatory response in mouse liver. Proc Natl Acad Sci U S A 91, 6196-6200.

Evans, M. J., von Hahn, T., Tscherne, D. M., Syder, A. J., Panis, M., Wolk, B., Hatziioannou, T., McKeating, J. A., Bieniasz, P. D., and Rice, C. M. (2007). Claudin-1 is a hepatitis C virus co-receptor required for a late step in entry. Nature 446, 801-805.

Fallaux, F. J., Bout, A., van der Velde, I., van den Wollenberg, D. J., Hehir, K. M., Keegan, J., Auger, C., Cramer, S. J., van Ormondt, H., van der Eb, A. J., et al. (1998). New helper cells and matched early region 1-deleted adenovirus vectors prevent generation of replication-competent adenoviruses. Hum Gene Ther 9, 1909-1917.

Fallaux, F. J., Kranenburg, O., Cramer, S. J., Houweling, A., Van Ormondt, H., Hoeben, R. C., and Van Der Eb, A. J. (1996). Characterization of 911: a new helper cell line for the titration and propagation of early region 1-deleted adenoviral vectors. Hum Gene Ther 7, 215-222.

# REFERENCES

Fallaux, F. J., van der Eb, A. J., and Hoeben, R. C. (1999). Who's afraid of replication-competent adenoviruses? Gene Ther 6, 709-712.

Fasano, A., Baudry, B., Pumplin, D. W., Wasserman, S. S., Tall, B. D., Ketley, J. M., and Kaper, J. B. (1991). Vibrio cholerae produces a second enterotoxin, which affects intestinal tight junctions. Proceedings of the National Academy of Sciences of the United States of America 88, 5242-5246.

Fasano, A., Fiorentini, C., Donelli, G., Uzzau, S., Kaper, J. B., Margaretten, K., Ding, X., Guandalini, S., Comstock, L., and Goldblum, S. E. (1995). Zonula occludens toxin modulates tight junctions through protein kinase C-dependent actin reorganization, in vitro. J Clin Invest 96, 710-720.

Fechner, H., Haack, A., Wang, H., Wang, X., Eizema, K., Pauschinger, M., Schoemaker, R., Veghel, R., Houtsmuller, A., Schultheiss, H. P., *et al.* (1999). Expression of coxsackie adenovirus receptor and alphav-integrin does not correlate with adenovector targeting in vivo indicating anatomical vector barriers. Gene Ther 6, 1520-1535.

Feeley, K. M., and Wells, M. (2001). Precursor lesions of ovarian epithelial malignancy. Histopathology 38, 87-95.

Fender, P., Boussaid, A., Mezin, P., and Chroboczek, J. (2005). Synthesis, cellular localization, and quantification of penton-dodecahedron in serotype 3 adenovirus-infected cells. Virology 340, 167-173.

Fender, P., Ruigrok, R. W., Gout, E., Buffet, S., and Chroboczek, J. (1997). Adenovirus dodecahedron, a new vector for human gene transfer. Nat Biotechnol 15, 52-56.

Fender, P., Schoehn, G., Foucaud-Gamen, J., Gout, E., Garcel, A., Drouet, E., and Chroboczek, J. (2003). Adenovirus dodecahedron allows large multimeric protein transduction in human cells. J Virol 77, 4960-4964.

Fender, P., Schoehn, G., Perron-Sierra, F., Tucker, G. C., and Lortat-Jacob, H. (2008). Adenovirus dodecahedron cell attachment and entry are mediated by heparan sulfate and integrins and vary along the cell cycle. Virology 371, 155-164.

Filipowicz, W., Bhattacharyya, S. N., and Sonenberg, N. (2008). Mechanisms of post-transcriptional regulation by microRNAs: are the answers in sight? Nat Rev Genet 9, 102-114.

Fisher, K. J., Choi, H., Burda, J., Chen, S. J., and Wilson, J. M. (1996). Recombinant adenovirus deleted of all viral genes for gene therapy of cystic fibrosis. Virology 217, 11-22.

Fogh, J., Fogh, J. M., and Orfeo, T. (1977). One hundred and twenty-seven cultured human tumor cell lines producing tumors in nude mice. J Natl Cancer Inst 59, 221-226.

Folkman, J., and Shing, Y. (1992). Angiogenesis. J Biol Chem 267, 10931-10934.

Fox, D. T., and Peifer, M. (2007). Cell adhesion: separation of p120's powers? Curr Biol 17, R24-27.

Franci, C., Takkunen, M., Dave, N., Alameda, F., Gomez, S., Rodriguez, R., Escriva, M., Montserrat-Sentis, B., Baro, T., Garrido, M., *et al.* (2006). Expression of Snail protein in tumor-stroma interface. Oncogene 25, 5134-5144.

Fredman, J. N., and Engler, J. A. (1993). Adenovirus precursor to terminal protein interacts with the nuclear matrix in vivo and in vitro. J Virol 67, 3384-3395.

Freimuth, P., Springer, K., Berard, C., Hainfeld, J., Bewley, M., and Flanagan, J. (1999). Coxsackievirus and adenovirus receptor amino-terminal immunoglobulin V-related domain binds adenovirus type 2 and fiber knob from adenovirus type 12. J Virol 73, 1392-1398.

# REFERENCES

Frisch, S. M. (1997). The epithelial cell default-phenotype hypothesis and its implications for cancer. Bioessays 19, 705-709.

Fueyo, J., Gomez-Manzano, C., Alemany, R., Lee, P. S., McDonnell, T. J., Mitlianga, P., Shi, Y. X., Levin, V. A., Yung, W. K., and Kyritsis, A. P. (2000). A mutant oncolytic adenovirus targeting the Rb pathway produces anti-glioma effect in vivo. Oncogene 19, 2-12.

Fujita, N., Jaye, D. L., Kajita, M., Geigerman, C., Moreno, C. S., and Wade, P. A. (2003). MTA3, a Mi-2/NuRD complex subunit, regulates an invasive growth pathway in breast cancer. Cell 113, 207-219.

Fujiwara, T., Urata, Y., and Tanaka, N. (2007). Telomerase-specific oncolytic virotherapy for human cancer with the hTERT promoter. Curr Cancer Drug Targets 7, 191-201.

Fukumura, D., and Jain, R. K. (2007). Tumor microenvironment abnormalities: causes, consequences, and strategies to normalize. J Cell Biochem 101, 937-949.

Funayama, N., Sato, Y., Matsumoto, K., Ogura, T., and Takahashi, Y. (1999). Coelom formation: binary decision of the lateral plate mesoderm is controlled by the ectoderm. Development 126, 4129-4138.

Furuse, M., Fujita, K., Hiiragi, T., Fujimoto, K., and Tsukita, S. (1998). Claudin-1 and -2: novel integral membrane proteins localizing at tight junctions with no sequence similarity to occludin. J Cell Biol 141, 1539-1550.

Furuse, M., Furuse, K., Sasaki, H., and Tsukita, S. (2001). Conversion of zonulae occludentes from tight to leaky strand type by introducing claudin-2 into Madin-Darby canine kidney I cells. J Cell Biol 153, 263-272.

Furuse, M., Hirase, T., Itoh, M., Nagafuchi, A., Yonemura, S., and Tsukita, S. (1993). Occludin: a novel integral membrane protein localizing at tight junctions. J Cell Biol 123, 1777-1788.

Ganly, I., Kirn, D., Eckhardt, G., Rodriguez, G. I., Soutar, D. S., Otto, R., Robertson, A. G., Park, O., Gulley, M. L., Heise, C., et al. (2000). A phase I study of Onyx-015, an E1B attenuated adenovirus, administered intratumorally to patients with recurrent head and neck cancer. Clin Cancer Res 6, 798-806.

Ganz, T. (2003). Defensins: antimicrobial peptides of innate immunity. Nat Rev Immunol 3, 710-720.

Gao, Y., Dickerson, J. B., Guo, F., Zheng, J., and Zheng, Y. (2004). Rational design and characterization of a Rac GTPase-specific small molecule inhibitor. Proc Natl Acad Sci U S A 101, 7618-7623.

Garzon, R., Fabbri, M., Cimmino, A., Calin, G. A., and Croce, C. M. (2006). MicroRNA expression and function in cancer. Trends Mol Med 12, 580-587.

Gatenby, R. A., and Gillies, R. J. (2004). Why do cancers have high aerobic glycolysis? Nat Rev Cancer 4, 891-899.

Genth, H., Gerhard, R., Maeda, A., Amano, M., Kaibuchi, K., Aktories, K., and Just, I. (2003). Entrapment of Rho ADP-ribosylated by Clostridium botulinum C3 exoenzyme in the Rho-guanine nucleotide dissociation inhibitor-1 complex. J Biol Chem 278, 28523-28527.

Geraghty, R. J., Krummenacher, C., Cohen, G. H., Eisenberg, R. J., and Spear, P. G. (1998). Entry of alphaherpesviruses mediated by poliovirus receptor-related protein 1 and poliovirus receptor. Science 280, 1618-1620.

Gershengorn, M. C., Hardikar, A. A., Wei, C., Geras-Raaka, E., Marcus-Samuels, B., and Raaka, B. M. (2004). Epithelial-to-mesenchymal transition generates proliferative human islet precursor cells. Science 306, 2261-2264.

# REFERENCES

Giroldi, L. A., Bringuier, P. P., de Weijert, M., Jansen, C., van Bokhoven, A., and Schalken, J. A. (1997). Role of E boxes in the repression of E-cadherin expression. Biochem Biophys Res Commun 241, 453-458.

Goosney, D. L., and Nemerow, G. R. (2003). Adenovirus infection: taking the back roads to viral entry. Curr Biol 13, R99-R100.

Gordon, Y. J., Huang, L. C., Romanowski, E. G., Yates, K. A., Proske, R. J., and McDermott, A. M. (2005). Human cathelicidin (LL-37), a multifunctional peptide, is expressed by ocular surface epithelia and has potent antibacterial and antiviral activity. Curr Eye Res 30, 385-394.

Gorlach, A., Herter, P., Hentschel, H., Frosch, P. J., and Acker, H. (1994). Effects of nIFN beta and rIFN gamma on growth and morphology of two human melanoma cell lines: comparison between two- and three-dimensional culture. Int J Cancer 56, 249-254.

Gort, E. H., Groot, A. J., van der Wall, E., van Diest, P. J., and Vooijs, M. A. (2008). Hypoxic regulation of metastasis via hypoxia-inducible factors. Curr Mol Med 8, 60-67.

Gow, A., Southwood, C. M., Li, J. S., Pariali, M., Riordan, G. P., Brodie, S. E., Danias, J., Bronstein, J. M., Kachar, B., and Lazzarini, R. A. (1999). CNS myelin and sertoli cell tight junction strands are absent in Osp/claudin-11 null mice. Cell 99, 649-659.

Grady, W. M., Willis, J., Guilford, P. J., Dunbier, A. K., Toro, T. T., Lynch, H., Wiesner, G., Ferguson, K., Eng, C., Park, J. G., et al. (2000). Methylation of the CDH1 promoter as the second genetic hit in hereditary diffuse gastric cancer. Nat Genet 26, 16-17.

Graham, F. L., Smiley, J., Russell, W. C., and Nairn, R. (1977). Characteristics of a human cell line transformed by DNA from human adenovirus type 5. J Gen Virol 36, 59-74.

Greber, U. F., Willetts, M., Webster, P., and Helenius, A. (1993). Stepwise dismantling of adenovirus 2 during entry into cells. Cell 75, 477-486.

Greenblatt, M. S., Bennett, W. P., Hollstein, M., and Harris, C. C. (1994). Mutations in the p53 tumor suppressor gene: clues to cancer etiology and molecular pathogenesis. Cancer Res 54, 4855-4878.

Gregoire, L., Munkarah, A., Rabah, R., Morris, R. T., and Lancaster, W. D. (1998). Organotypic culture of human ovarian surface epithelial cells: a potential model for ovarian carcinogenesis. In Vitro Cell Dev Biol Anim 34, 636-639.

Gregory, P. A., Bert, A. G., Paterson, E. L., Barry, S. C., Tsykin, A., Farshid, G., Vadas, M. A., Khew-Goodall, Y., and Goodall, G. J. (2008a). The miR-200 family and miR-205 regulate epithelial to mesenchymal transition by targeting ZEB1 and SIP1. Nat Cell Biol 10, 593-601.

Gregory, P. A., Bracken, C. P., Bert, A. G., and Goodall, G. J. (2008b). MicroRNAs as regulators of epithelial-mesenchymal transition. Cell Cycle 7, 3112-3118.

Gross, S. (1971). Measles and leukaemia. Lancet 1, 397-398.

Gumbiner, B., Stevenson, B., and Grimaldi, A. (1988). The role of the cell adhesion molecule uvomorulin in the formation and maintenance of the epithelial junctional complex. J Cell Biol 107, 1575-1587.

Gumbiner, B. M. (2005). Regulation of cadherin-mediated adhesion in morphogenesis. Nat Rev Mol Cell Biol 6, 622-634.

Hajra, K. M., and Fearon, E. R. (2002). Cadherin and catenin alterations in human cancer. Genes Chromosomes Cancer 34, 255-268.

# REFERENCES

Hajra, K. M., Ji, X., and Fearon, E. R. (1999). Extinction of E-cadherin expression in breast cancer via a dominant repression pathway acting on proximal promoter elements. Oncogene *18*, 7274-7279.

Halbleib, J. M., and Nelson, W. J. (2006). Cadherins in development: cell adhesion, sorting, and tissue morphogenesis. Genes Dev *20*, 3199-3214.

Hallenbeck, P. L., Chang, Y. N., Hay, C., Golightly, D., Stewart, D., Lin, J., Phipps, S., and Chiang, Y. L. (1999). A novel tumor-specific replication-restricted adenoviral vector for gene therapy of hepatocellular carcinoma. Hum Gene Ther *10*, 1721-1733.

Hamid, O., Varterasian, M. L., Wadler, S., Hecht, J. R., Benson, A., 3rd, Galanis, E., Uprichard, M., Omer, C., Bycott, P., Hackman, R. C., and Shields, A. F. (2003). Phase II trial of intravenous CI-1042 in patients with metastatic colorectal cancer. J Clin Oncol *21*, 1498-1504.

Han, Z., Xu, G., Zhou, J., Xing, H., Wang, S., Wu, M., Zhang, Y., Lu, Y., and Ma, D. (2005). Inhibition of motile and invasive properties of ovarian cancer cells by ASODN against Rho-associated protein kinase. Cell Commun Adhes *12*, 59-69.

Hanahan, D., and Weinberg, R. A. (2000). The hallmarks of cancer. Cell *100*, 57-70.

Harada, J. N., and Berk, A. J. (1999). p53-Independent and -dependent requirements for E1B-55K in adenovirus type 5 replication. J Virol *73*, 5333-5344.

Hartmann, L. C., Lu, K. H., Linette, G. P., Cliby, W. A., Kalli, K. R., Gershenson, D., Bast, R. C., Stec, J., Iartchouk, N., Smith, D. I., *et al.* (2005). Gene expression profiles predict early relapse in ovarian cancer after platinum-paclitaxel chemotherapy. Clin Cancer Res *11*, 2149-2155.

Hatefi, A., Cappello, J., and Ghandehari, H. (2007). Adenoviral gene delivery to solid tumors by recombinant silk-elastinlike protein polymers. Pharmaceutical research *24*, 773-779.

Hawkins, P. G., and Morris, K. V. (2008). RNA and transcriptional modulation of gene expression. Cell Cycle *7*, 602-607.

Heinzelmann-Schwarz, V. A., Gardiner-Garden, M., Henshall, S. M., Scurry, J., Scolyer, R. A., Davies, M. J., Heinzelmann, M., Kalish, L. H., Bali, A., Kench, J. G., *et al.* (2004). Overexpression of the cell adhesion molecules DDR1, Claudin 3, and Ep-CAM in metaplastic ovarian epithelium and ovarian cancer. Clin Cancer Res *10*, 4427-4436.

Heise, C., and Kirn, D. H. (2000). Replication-selective adenoviruses as oncolytic agents. J Clin Invest *105*, 847-851.

Hermiston, T. (2006). A demand for next-generation oncolytic adenoviruses. Curr Opin Mol Ther *8*, 322-330.

Hoffmann, D., Bayer, W., Heim, A., Potthoff, A., Nettelbeck, D. M., and Wildner, O. (2008). Evaluation of twenty-one human adenovirus types and one infectivity-enhanced adenovirus for the treatment of malignant melanoma. J Invest Dermatol *128*, 988-998.

Hoffmann, D., and Wildner, O. (2006). Efficient generation of double heterologous promoter controlled oncolytic adenovirus vectors by a single homologous recombination step in Escherichia coli. BMC Biotechnol *6*.

Holm, P. S., Bergmann, S., Jurchott, K., Lage, H., Brand, K., Ladhoff, A., Mantwill, K., Curiel, D. T., Dobbelstein, M., Dietel, M., *et al.* (2002). YB-1 relocates to the nucleus in adenovirus-infected cells and facilitates viral replication by inducing E2 gene expression through the E2 late promoter. J Biol Chem *277*, 10427-10434.

## References

Howe, J. A., Demers, G. W., Johnson, D. E., Neugebauer, S. E., Perry, S. T., Vaillancourt, M. T., and Faha, B. (2000). Evaluation of E1-mutant adenoviruses as conditionally replicating agents for cancer therapy. Mol Ther 2, 485-495.

Huang, Y. H., Bao, Y., Peng, W., Goldberg, M., Love, K., Bumcrot, D. A., Cole, G., Langer, R., Anderson, D. G., and Sawicki, J. A. (2009). Claudin-3 gene silencing with siRNA suppresses ovarian tumor growth and metastasis. Proceedings of the National Academy of Sciences of the United States of America 106, 3426-3430.

Huber, A. H., Stewart, D. B., Laurents, D. V., Nelson, W. J., and Weis, W. I. (2001). The cadherin cytoplasmic domain is unstructured in the absence of beta-catenin. A possible mechanism for regulating cadherin turnover. J Biol Chem 276, 12301-12309.

Huber, A. H., and Weis, W. I. (2001). The structure of the beta-catenin/E-cadherin complex and the molecular basis of diverse ligand recognition by beta-catenin. Cell 105, 391-402.

Hudson, L. G., Zeineldin, R., and Stack, M. S. (2008). Phenotypic plasticity of neoplastic ovarian epithelium: unique cadherin profiles in tumor progression. Clin Exp Metastasis 25, 643-655.

Huebner, R. J., Rowe, W. P., Schatten, W. E., Smith, R. R., and Thomas, L. B. (1956). Studies on the use of viruses in the treatment of carcinoma of the cervix. Cancer 9, 1211-1218.

Hynes, R. O. (2002). Integrins: bidirectional, allosteric signaling machines. Cell 110, 673-687.

Ikenouchi, J., Furuse, M., Furuse, K., Sasaki, H., and Tsukita, S. (2005). Tricellulin constitutes a novel barrier at tricellular contacts of epithelial cells. J Cell Biol 171, 939-945.

Ikenouchi, J., Matsuda, M., Furuse, M., and Tsukita, S. (2003). Regulation of tight junctions during the epithelium-mesenchyme transition: direct repression of the gene expression of claudins/occludin by Snail. J Cell Sci 116, 1959-1967.

Ikenoya, M., Hidaka, H., Hosoya, T., Suzuki, M., Yamamoto, N., and Sasaki, Y. (2002). Inhibition of rho-kinase-induced myristoylated alanine-rich C kinase substrate (MARCKS) phosphorylation in human neuronal cells by H-1152, a novel and specific Rho-kinase inhibitor. J Neurochem 81, 9-16.

Imai, T., Horiuchi, A., Shiozawa, T., Osada, R., Kikuchi, N., Ohira, S., Oka, K., and Konishi, I. (2004). Elevated expression of E-cadherin and alpha-, beta-, and gamma-catenins in metastatic lesions compared with primary epithelial ovarian carcinomas. Hum Pathol 35, 1469-1476.

Iyer, A. K., Khaled, G., Fang, J., and Maeda, H. (2006). Exploiting the enhanced permeability and retention effect for tumor targeting. Drug Discov Today 11, 812-818.

Jain, R. K. (1990). Vascular and interstitial barriers to delivery of therapeutic agents in tumors. Cancer Metastasis Rev 9, 253-266.

Jain, R. K. (1997). Delivery of molecular and cellular medicine to solid tumors. Adv Drug Deliv Rev 26, 71-90.

Jain, R. K. (1998). Delivery of molecular and cellular medicine to solid tumors. J Control Release 53, 49-67.

Jain, R. K. (2008). Lessons from multidisciplinary translational trials on anti-angiogenic therapy of cancer. Nat Rev Cancer 8, 309-316.

Jain, R. K., Tong, R. T., and Munn, L. L. (2007). Effect of vascular normalization by antiangiogenic therapy on interstitial hypertension, peritumor edema, and lymphatic metastasis: insights from a mathematical model. Cancer Res 67, 2729-2735.

# REFERENCES

James, D., Levine, A. J., Besser, D., and Hemmati-Brivanlou, A. (2005). TGFbeta/activin/nodal signaling is necessary for the maintenance of pluripotency in human embryonic stem cells. Development *132*, 1273-1282.

Janmey, P. A., Euteneuer, U., Traub, P., and Schliwa, M. (1991). Viscoelastic properties of vimentin compared with other filamentous biopolymer networks. J Cell Biol *113*, 155-160.

Jemal, A., Siegel, R., Ward, E., Hao, Y., Xu, J., Murray, T., and Thun, M. J. (2008). Cancer statistics, 2008. CA Cancer J Clin *58*, 71-96.

Jiang, H., Wang, Z., Serra, D., Frank, M. M., and Amalfitano, A. (2004). Recombinant adenovirus vectors activate the alternative complement pathway, leading to the binding of human complement protein C3 independent of anti-ad antibodies. Mol Ther *10*, 1140-1142.

Jounaidi, Y., Doloff, J. C., and Waxman, D. J. (2007). Conditionally replicating adenoviruses for cancer treatment. Curr Cancer Drug Targets *7*, 285-301.

Kalyuzhniy, O., Di Paolo, N. C., Silvestry, M., Hofherr, S. E., Barry, M. A., Stewart, P. L., and Shayakhmetov, D. M. (2008). Adenovirus serotype 5 hexon is critical for virus infection of hepatocytes in vivo. Proc Natl Acad Sci U S A *105*, 5483-5488.

Kanerva, A., Wang, M., Bauerschmitz, G. J., Lam, J. T., Desmond, R. A., Bhoola, S. M., Barnes, M. N., Alvarez, R. D., Siegal, G. P., Curiel, D. T., and Hemminki, A. (2002). Gene transfer to ovarian cancer versus normal tissues with fiber-modified adenoviruses. Mol Ther *5*, 695-704.

Keirsebilck, A., Bonne, S., Staes, K., van Hengel, J., Nollet, F., Reynolds, A., and van Roy, F. (1998). Molecular cloning of the human p120ctn catenin gene (CTNND1): expression of multiple alternatively spliced isoforms. Genomics *50*, 129-146.

Ketner, G., Spencer, F., Tugendreich, S., Connelly, C., and Hieter, P. (1994). Efficient manipulation of the human adenovirus genome as an infectious yeast artificial chromosome clone. Proc Natl Acad Sci U S A *91*, 6186-6190.

Khatri, P., Desai, V., Tarca, A. L., Sellamuthu, S., Wildman, D. E., Romero, R., and Draghici, S. (2006). New Onto-Tools: Promoter-Express, nsSNPCounter and Onto-Translate. Nucleic acids research *34*, W626-631.

Khuri, F. R., Nemunaitis, J., Ganly, I., Arseneau, J., Tannock, I. F., Romel, L., Gore, M., Ironside, J., MacDougall, R. H., Heise, C., et al. (2000). a controlled trial of intratumoral ONYX-015, a selectively-replicating adenovirus, in combination with cisplatin and 5-fluorouracil in patients with recurrent head and neck cancer. Nat Med *6*, 879-885.

Kiang, A., Hartman, Z. C., Everett, R. S., Serra, D., Jiang, H., Frank, M. M., and Amalfitano, A. (2006). Multiple innate inflammatory responses induced after systemic adenovirus vector delivery depend on a functional complement system. Mol Ther *14*, 588-598.

Kirby, T. O., Rivera, A., Rein, D., Wang, M., Ulasov, I., Breidenbach, M., Kataram, M., Contreras, J. L., Krumdieck, C., Yamamoto, M., et al. (2004). A novel ex vivo model system for evaluation of conditionally replicative adenoviruses therapeutic efficacy and toxicity. Clin Cancer Res *10*, 8697-8703.

Kochanek, S., Clemens, P. R., Mitani, K., Chen, H. H., Chan, S., and Caskey, C. T. (1996). A new adenoviral vector: Replacement of all viral coding sequences with 28 kb of DNA independently expressing both full-length dystrophin and beta-galactosidase. Proc Natl Acad Sci U S A *93*, 5731-5736.

Koizume, S., Tachibana, K., Sekiya, T., Hirohashi, S., and Shiraishi, M. (2002). Heterogeneity in the modification and involvement of chromatin components of the CpG island of the silenced human CDH1 gene in cancer cells. Nucleic Acids Res *30*, 4770-4780.

# References

Kollmar, R., Nakamura, S. K., Kappler, J. A., and Hudspeth, A. J. (2001). Expression and phylogeny of claudins in vertebrate primordia. Proc Natl Acad Sci U S A *98*, 10196-10201.

Kurihara, T., Brough, D. E., Kovesdi, I., and Kufe, D. W. (2000). Selectivity of a replication-competent adenovirus for human breast carcinoma cells expressing the MUC1 antigen. J Clin Invest *106*, 763-771.

Kuriyama, N., Kuriyama, H., Julin, C. M., Lamborn, K., and Israel, M. A. (2000). Pretreatment with protease is a useful experimental strategy for enhancing adenovirus-mediated cancer gene therapy. Hum Gene Ther *11*, 2219-2230.

LaGamba, D., Nawshad, A., and Hay, E. D. (2005). Microarray analysis of gene expression during epithelial-mesenchymal transformation. Dev Dyn *234*, 132-142.

Lee, Y. S., Kim, J. H., Choi, K. J., Choi, I. K., Kim, H., Cho, S., Cho, B. C., and Yun, C. O. (2006). Enhanced antitumor effect of oncolytic adenovirus expressing interleukin-12 and B7-1 in an immunocompetent murine model. Clin Cancer Res *12*, 5859-5868.

Lejmi, E., Leconte, L., Pedron-Mazoyer, S., Ropert, S., Raoul, W., Lavalette, S., Bouras, I., Feron, J. G., Maitre-Boube, M., Assayag, F., *et al.* (2008). Netrin-4 inhibits angiogenesis via binding to neogenin and recruitment of Unc5B. Proc Natl Acad Sci U S A *105*, 12491-12496.

Leopold, P. L., Kreitzer, G., Miyazawa, N., Rempel, S., Pfister, K. K., Rodriguez-Boulan, E., and Crystal, R. G. (2000). Dynein- and microtubule-mediated translocation of adenovirus serotype 5 occurs after endosomal lysis. Hum Gene Ther *11*, 151-165.

Leppard, K. N. (1997). E4 gene function in adenovirus, adenovirus vector and adeno-associated virus infections. J Gen Virol *78 ( Pt 9)*, 2131-2138.

Leptin, M. (1991). twist and snail as positive and negative regulators during Drosophila mesoderm development. Genes Dev *5*, 1568-1576.

Li, E., Stupack, D., Bokoch, G. M., and Nemerow, G. R. (1998a). Adenovirus endocytosis requires actin cytoskeleton reorganization mediated by Rho family GTPases. J Virol *72*, 8806-8812.

Li, E., Stupack, D., Klemke, R., Cheresh, D. A., and Nemerow, G. R. (1998b). Adenovirus endocytosis via alpha(v) integrins requires phosphoinositide-3-OH kinase. J Virol *72*, 2055-2061.

Li, Z., Liu, Y., Tuve, S., Xun, Y., Fan, X., Min, L., Feng, Q., Kiviat, N., Kiem, H. P., Disis, M. L., and Lieber, A. (2009). Towards a stem cell gene therapy for breast cancer. Blood.

Li, Z. Y., Ni, S., Yang, X., Kiviat, N., and Lieber, A. (2004). Xenograft models for liver metastasis: Relationship between tumor morphology and adenovirus vector transduction. Mol Ther *9*, 650-657.

Lickert, H., Bauer, A., Kemler, R., and Stappert, J. (2000). Casein kinase II phosphorylation of E-cadherin increases E-cadherin/beta-catenin interaction and strengthens cell-cell adhesion. J Biol Chem *275*, 5090-5095.

Lilien, J., Balsamo, J., Arregui, C., and Xu, G. (2002). Turn-off, drop-out: functional state switching of cadherins. Dev Dyn *224*, 18-29.

Liu, Y., Nusrat, A., Schnell, F. J., Reaves, T. A., Walsh, S., Pochet, M., and Parkos, C. A. (2000). Human junction adhesion molecule regulates tight junction resealing in epithelia. J Cell Sci *113 ( Pt 13)*, 2363-2374.

Liu, Y., Wang, H., Yumul, R., Gao, W., Gambotto, A., Morita, T., Baker, A., Shayakhmetov, D., and Lieber, A. (2009). Transduction of liver metastases after intravenous injection of Ad5/35 or Ad35 vectors with and without factor X-binding protein pretreatment. Hum Gene Ther *20*, 621-629.

# References

Locascio, A., and Nieto, M. A. (2001). Cell movements during vertebrate development: integrated tissue behaviour versus individual cell migration. Curr Opin Genet Dev *11*, 464-469.

Loeb, L. A., Bielas, J. H., and Beckman, R. A. (2008). Cancers exhibit a mutator phenotype: clinical implications. Cancer Res *68*, 3551-3557; discussion 3557.

Ma, L., Teruya-Feldstein, J., and Weinberg, R. A. (2007). Tumour invasion and metastasis initiated by microRNA-10b in breast cancer. Nature *449*, 682-688.

Madara, J. L. (1998). Regulation of the movement of solutes across tight junctions. Annu Rev Physiol *60*, 143-159.

Maeda, H., Wu, J., Sawa, T., Matsumura, Y., and Hori, K. (2000). Tumor vascular permeability and the EPR effect in macromolecular therapeutics: a review. J Control Release *65*, 271-284.

Mani, S. A., Guo, W., Liao, M. J., Eaton, E. N., Ayyanan, A., Zhou, A. Y., Brooks, M., Reinhard, F., Zhang, C. C., Shipitsin, M., *et al.* (2008). The epithelial-mesenchymal transition generates cells with properties of stem cells. Cell *133*, 704-715.

Martinez-Estrada, O. M., Culleres, A., Soriano, F. X., Peinado, H., Bolos, V., Martinez, F. O., Reina, M., Cano, A., Fabre, M., and Vilaro, S. (2006). The transcription factors Slug and Snail act as repressors of Claudin-1 expression in epithelial cells. Biochem J *394*, 449-457.

Martuza, R. L., Malick, A., Markert, J. M., Ruffner, K. L., and Coen, D. M. (1991). Experimental therapy of human glioma by means of a genetically engineered virus mutant. Science *252*, 854-856.

Massague, J. (2008). TGFbeta in Cancer. Cell *134*, 215-230.

Matsunaga, M., Hatta, K., Nagafuchi, A., and Takeichi, M. (1988). Guidance of optic nerve fibres by N-cadherin adhesion molecules. Nature *334*, 62-64.

Maxwell, S. A., and Davis, G. E. (2000). Differential gene expression in p53-mediated apoptosis-resistant vs. apoptosis-sensitive tumor cell lines. Proc Natl Acad Sci U S A *97*, 13009-13014.

McGrory, W. J., Bautista, D. S., and Graham, F. L. (1988). A simple technique for the rescue of early region I mutations into infectious human adenovirus type 5. Virology *163*, 614-617.

Medici, D., Hay, E. D., and Goodenough, D. A. (2006). Cooperation between snail and LEF-1 transcription factors is essential for TGF-beta1-induced epithelial-mesenchymal transition. Mol Biol Cell *17*, 1871-1879.

Menozzi, F. D., Reddy, V. M., Cayet, D., Raze, D., Debrie, A. S., Dehouck, M. P., Cecchelli, R., and Locht, C. (2006). Mycobacterium tuberculosis heparin-binding haemagglutinin adhesin (HBHA) triggers receptor-mediated transcytosis without altering the integrity of tight junctions. Microbes and infection / Institut Pasteur *8*, 1-9.

Mi, J., Li, Z. Y., Ni, S., Steinwaerder, D., and Lieber, A. (2001). Induced apoptosis supports spread of adenovirus vectors in tumors. Hum Gene Ther *12*, 1343-1352.

Milosevic, M. F., Fyles, A. W., Wong, R., Pintilie, M., Kavanagh, M. C., Levin, W., Manchul, L. A., Keane, T. J., and Hill, R. P. (1998). Interstitial fluid pressure in cervical carcinoma: within tumor heterogeneity, and relation to oxygen tension. Cancer *82*, 2418-2426.

Mittereder, N., March, K. L., and Trapnell, B. C. (1996). Evaluation of the concentration and bioactivity of adenovirus vectors for gene therapy. J Virol *70*, 7498-7509.

# REFERENCES

Miyamoto, T., Morita, K., Takemoto, D., Takeuchi, K., Kitano, Y., Miyakawa, T., Nakayama, K., Okamura, Y., Sasaki, H., Miyachi, Y., et al. (2005). Tight junctions in Schwann cells of peripheral myelinated axons: a lesson from claudin-19-deficient mice. J Cell Biol *169*, 527-538.

Mo, Y. Y., and Reynolds, A. B. (1996). Identification of murine p120 isoforms and heterogeneous expression of p120cas isoforms in human tumor cell lines. Cancer Res *56*, 2633-2640.

Mok, W., Boucher, Y., and Jain, R. K. (2007). Matrix metalloproteinases-1 and -8 improve the distribution and efficacy of an oncolytic virus. Cancer Res *67*, 10664-10668.

Moon, R. T., Bowerman, B., Boutros, M., and Perrimon, N. (2002). The promise and perils of Wnt signaling through beta-catenin. Science *296*, 1644-1646.

Morel, A. P., Lievre, M., Thomas, C., Hinkal, G., Ansieau, S., and Puisieux, A. (2008). Generation of breast cancer stem cells through epithelial-mesenchymal transition. PLoS ONE *3*, e2888.

Moskalenko, M., Chen, L., van Roey, M., Donahue, B. A., Snyder, R. O., McArthur, J. G., and Patel, S. D. (2000). Epitope mapping of human anti-adeno-associated virus type 2 neutralizing antibodies: implications for gene therapy and virus structure. J Virol *74*, 1761-1766.

Mulvihill, S., Warren, R., Venook, A., Adler, A., Randlev, B., Heise, C., and Kirn, D. (2001). Safety and feasibility of injection with an E1B-55 kDa gene-deleted, replication-selective adenovirus (ONYX-015) into primary carcinomas of the pancreas: a phase I trial. Gene Ther *8*, 308-315.

Murakami, S., Sakurai, F., Kawabata, K., Okada, N., Fujita, T., Yamamoto, A., Hayakawa, T., and Mizuguchi, H. (2007). Interaction of penton base Arg-Gly-Asp motifs with integrins is crucial for adenovirus serotype 35 vector transduction in human hematopoietic cells. Gene Ther *14*, 1525-1533.

Muresan, Z., Paul, D. L., and Goodenough, D. A. (2000). Occludin 1B, a variant of the tight junction protein occludin. Mol Biol Cell *11*, 627-634.

Nagano, S., Perentes, J. Y., Jain, R. K., and Boucher, Y. (2008). Cancer cell death enhances the penetration and efficacy of oncolytic herpes simplex virus in tumors. Cancer Res *68*, 3795-3802.

Naora, H. (2007). The heterogeneity of epithelial ovarian cancers: reconciling old and new paradigms. Expert Rev Mol Med *9*, 1-12.

Nemerow, G. R., and Stewart, P. L. (1999). Role of alpha(v) integrins in adenovirus cell entry and gene delivery. Microbiol Mol Biol Rev *63*, 725-734.

Nemunaitis, J., Cunningham, C., Buchanan, A., Blackburn, A., Edelman, G., Maples, P., Netto, G., Tong, A., Randlev, B., Olson, S., and Kirn, D. (2001a). Intravenous infusion of a replication-selective adenovirus (ONYX-015) in cancer patients: safety, feasibility and biological activity. Gene Ther *8*, 746-759.

Nemunaitis, J., Khuri, F., Ganly, I., Arseneau, J., Posner, M., Vokes, E., Kuhn, J., McCarty, T., Landers, S., Blackburn, A., et al. (2001b). Phase II trial of intratumoral administration of ONYX-015, a replication-selective adenovirus, in patients with refractory head and neck cancer. J Clin Oncol *19*, 289-298.

Nettelbeck, D. M., Rivera, A. A., Davydova, J., Dieckmann, D., Yamamoto, M., and Curiel, D. T. (2003). Cyclooxygenase-2 promoter for tumour-specific targeting of adenoviral vectors to melanoma. Melanoma Res *13*, 287-292.

Nieto, M. A. (2002). The snail superfamily of zinc-finger transcription factors. Nat Rev Mol Cell Biol *3*, 155-166.

Nollet, F., Kools, P., and van Roy, F. (2000). Phylogenetic analysis of the cadherin superfamily allows identification of six major subfamilies besides several solitary members. J Mol Biol *299*, 551-572.

# References

Nusrat, A., Brown, G. T., Tom, J., Drake, A., Bui, T. T., Quan, C., and Mrsny, R. J. (2005). Multiple protein interactions involving proposed extracellular loop domains of the tight junction protein occludin. Mol Biol Cell *16*, 1725-1734.

O'Malley, R. P., Mariano, T. M., Siekierka, J., and Mathews, M. B. (1986). A mechanism for the control of protein synthesis by adenovirus VA RNAI. Cell *44*, 391-400.

Ohkubo, T., and Ozawa, M. (2004). The transcription factor Snail downregulates the tight junction components independently of E-cadherin downregulation. J Cell Sci *117*, 1675-1685.

Oliveira, S. S., and Morgado-Diaz, J. A. (2007). Claudins: multifunctional players in epithelial tight junctions and their role in cancer. Cell Mol Life Sci *64*, 17-28.

Ozdamar, B., Bose, R., Barrios-Rodiles, M., Wang, H. R., Zhang, Y., and Wrana, J. L. (2005). Regulation of the polarity protein Par6 by TGFbeta receptors controls epithelial cell plasticity. Science *307*, 1603-1609.

Padera, T. P., Stoll, B. R., Tooredman, J. B., Capen, D., di Tomaso, E., and Jain, R. K. (2004). Pathology: cancer cells compress intratumour vessels. Nature *427*, 695.

Panorchan, P., Thompson, M. S., Davis, K. J., Tseng, Y., Konstantopoulos, K., and Wirtz, D. (2006). Single-molecule analysis of cadherin-mediated cell-cell adhesion. J Cell Sci *119*, 66-74.

Park, S. M., Gaur, A. B., Lengyel, E., and Peter, M. E. (2008). The miR-200 family determines the epithelial phenotype of cancer cells by targeting the E-cadherin repressors ZEB1 and ZEB2. Genes Dev *22*, 894-907.

Parks, R. J., Chen, L., Anton, M., Sankar, U., Rudnicki, M. A., and Graham, F. L. (1996). A helper-dependent adenovirus vector system: removal of helper virus by Cre-mediated excision of the viral packaging signal. Proc Natl Acad Sci U S A *93*, 13565-13570.

Patan, S., Munn, L. L., and Jain, R. K. (1996). Intussusceptive microvascular growth in a human colon adenocarcinoma xenograft: a novel mechanism of tumor angiogenesis. Microvasc Res *51*, 260-272.

Patel, I. S., Madan, P., Getsios, S., Bertrand, M. A., and MacCalman, C. D. (2003). Cadherin switching in ovarian cancer progression. Int J Cancer *106*, 172-177.

Patel, S. D., Ciatto, C., Chen, C. P., Bahna, F., Rajebhosale, M., Arkus, N., Schieren, I., Jessell, T. M., Honig, B., Price, S. R., and Shapiro, L. (2006). Type II cadherin ectodomain structures: implications for classical cadherin specificity. Cell *124*, 1255-1268.

Pawlak, G., and Helfman, D. M. (2002). Post-transcriptional down-regulation of ROCKI/Rho-kinase through an MEK-dependent pathway leads to cytoskeleton disruption in Ras-transformed fibroblasts. Mol Biol Cell *13*, 336-347.

Peinado, H., Ballestar, E., Esteller, M., and Cano, A. (2004). Snail mediates E-cadherin repression by the recruitment of the Sin3A/histone deacetylase 1 (HDAC1)/HDAC2 complex. Mol Cell Biol *24*, 306-319.

Peinado, H., Olmeda, D., and Cano, A. (2007). Snail, Zeb and bHLH factors in tumour progression: an alliance against the epithelial phenotype? Nat Rev Cancer *7*, 415-428.

Peralta Soler, A., Knudsen, K. A., Tecson-Miguel, A., McBrearty, F. X., Han, A. C., and Salazar, H. (1997). Expression of E-cadherin and N-cadherin in surface epithelial-stromal tumors of the ovary distinguishes mucinous from serous and endometrioid tumors. Hum Pathol *28*, 734-739.

Perez-Moreno, M., and Fuchs, E. (2006). Catenins: keeping cells from getting their signals crossed. Dev Cell *11*, 601-612.

# References

Perez-Moreno, M. A., Locascio, A., Rodrigo, I., Dhondt, G., Portillo, F., Nieto, M. A., and Cano, A. (2001). A new role for E12/E47 in the repression of E-cadherin expression and epithelial-mesenchymal transitions. J Biol Chem *276*, 27424-27431.

Perrais, M., Chen, X., Perez-Moreno, M., and Gumbiner, B. M. (2007). E-cadherin homophilic ligation inhibits cell growth and epidermal growth factor receptor signaling independently of other cell interactions. Mol Biol Cell *18*, 2013-2025.

Pipiya, T., Sauthoff, H., Huang, Y. Q., Chang, B., Cheng, J., Heitner, S., Chen, S., Rom, W. N., and Hay, J. G. (2005). Hypoxia reduces adenoviral replication in cancer cells by downregulation of viral protein expression. Gene Ther *12*, 911-917.

Pirone, D. M., Liu, W. F., Ruiz, S. A., Gao, L., Raghavan, S., Lemmon, C. A., Romer, L. H., and Chen, C. S. (2006). An inhibitory role for FAK in regulating proliferation: a link between limited adhesion and RhoA-ROCK signaling. J Cell Biol *174*, 277-288.

Pittenger, M. F., Mackay, A. M., Beck, S. C., Jaiswal, R. K., Douglas, R., Mosca, J. D., Moorman, M. A., Simonetti, D. W., Craig, S., and Marshak, D. R. (1999). Multilineage potential of adult human mesenchymal stem cells. Science (New York, NY *284*, 143-147.

Pitti, R. M., Marsters, S. A., Ruppert, S., Donahue, C. J., Moore, A., and Ashkenazi, A. (1996). Induction of apoptosis by Apo-2 ligand, a new member of the tumor necrosis factor cytokine family. J Biol Chem *271*, 12687-12690.

Pokutta, S., Herrenknecht, K., Kemler, R., and Engel, J. (1994). Conformational changes of the recombinant extracellular domain of E-cadherin upon calcium binding. Eur J Biochem *223*, 1019-1026.

Pokutta, S., and Weis, W. I. (2000). Structure of the dimerization and beta-catenin-binding region of alpha-catenin. Mol Cell *5*, 533-543.

Polyak, K., and Weinberg, R. A. (2009). Transitions between epithelial and mesenchymal states: acquisition of malignant and stem cell traits. Nat Rev Cancer *9*, 265-273.

Post, D. E., Devi, N. S., Li, Z., Brat, D. J., Kaur, B., Nicholson, A., Olson, J. J., Zhang, Z., and Van Meir, E. G. (2004). Cancer therapy with a replicating oncolytic adenovirus targeting the hypoxic microenvironment of tumors. Clin Cancer Res *10*, 8603-8612.

Post, D. E., Sandberg, E. M., Kyle, M. M., Devi, N. S., Brat, D. J., Xu, Z., Tighiouart, M., and Van Meir, E. G. (2007). Targeted cancer gene therapy using a hypoxia inducible factor dependent oncolytic adenovirus armed with interleukin-4. Cancer Res *67*, 6872-6881.

Post, D. E., and Van Meir, E. G. (2003). A novel hypoxia-inducible factor (HIF) activated oncolytic adenovirus for cancer therapy. Oncogene *22*, 2065-2072.

Postigo, A. A., Depp, J. L., Taylor, J. J., and Kroll, K. L. (2003). Regulation of Smad signaling through a differential recruitment of coactivators and corepressors by ZEB proteins. EMBO J *22*, 2453-2462.

Qian, X., Karpova, T., Sheppard, A. M., McNally, J., and Lowy, D. R. (2004). E-cadherin-mediated adhesion inhibits ligand-dependent activation of diverse receptor tyrosine kinases. EMBO J *23*, 1739-1748.

Raki, M., Kanerva, A., Ristimaki, A., Desmond, R. A., Chen, D. T., Ranki, T., Sarkioja, M., Kangasniemi, L., and Hemminki, A. (2005). Combination of gemcitabine and Ad5/3-Delta24, a tropism modified conditionally replicating adenovirus, for the treatment of ovarian cancer. Gene Ther *12*, 1198-1205.

Rangel, L. B., Agarwal, R., D'Souza, T., Pizer, E. S., Alo, P. L., Lancaster, W. D., Gregoire, L., Schwartz, D. R., Cho, K. R., and Morin, P. J. (2003). Tight junction proteins claudin-3 and claudin-4 are frequently overexpressed in ovarian cancer but not in ovarian cystadenomas. Clin Cancer Res *9*, 2567-2575.

# References

Rao, L., Debbas, M., Sabbatini, P., Hockenbery, D., Korsmeyer, S., and White, E. (1992). The adenovirus E1A proteins induce apoptosis, which is inhibited by the E1B 19-kDa and Bcl-2 proteins. Proc Natl Acad Sci U S A *89*, 7742-7746.

Reichard, K. W., Lorence, R. M., Cascino, C. J., Peeples, M. E., Walter, R. J., Fernando, M. B., Reyes, H. M., and Greager, J. A. (1992). Newcastle disease virus selectively kills human tumor cells. J Surg Res *52*, 448-453.

Rodriguez, R., Schuur, E. R., Lim, H. Y., Henderson, G. A., Simons, J. W., and Henderson, D. R. (1997). Prostate attenuated replication competent adenovirus (ARCA) CN706: a selective cytotoxic for prostate-specific antigen-positive prostate cancer cells. Cancer Res *57*, 2559-2563.

Roelvink, P. W., Mi Lee, G., Einfeld, D. A., Kovesdi, I., and Wickham, T. J. (1999). Identification of a conserved receptor-binding site on the fiber proteins of CAR-recognizing adenoviridae. Science *286*, 1568-1571.

Rogulski, K. R., Freytag, S. O., Zhang, K., Gilbert, J. D., Paielli, D. L., Kim, J. H., Heise, C. C., and Kirn, D. H. (2000). In vivo antitumor activity of ONYX-015 is influenced by p53 status and is augmented by radiotherapy. Cancer Res *60*, 1193-1196.

Rothen-Rutishauser, B., Riesen, F. K., Braun, A., Gunthert, M., and Wunderli-Allenspach, H. (2002). Dynamics of tight and adherens junctions under EGTA treatment. J Membr Biol *188*, 151-162.

Rowe, W. P., Huebner, R. J., Gilmore, L. K., Parrott, R. H., and Ward, T. G. (1953). Isolation of a cytopathogenic agent from human adenoids undergoing spontaneous degeneration in tissue culture. Proc Soc Exp Biol Med *84*, 570-573.

Sabbah, M., Emami, S., Redeuilh, G., Julien, S., Prevost, G., Zimber, A., Ouelaa, R., Bracke, M., De Wever, O., and Gespach, C. (2008). Molecular signature and therapeutic perspective of the epithelial-to-mesenchymal transitions in epithelial cancers. Drug Resist Updat *11*, 123-151.

Saeed, A. I., Sharov, V., White, J., Li, J., Liang, W., Bhagabati, N., Braisted, J., Klapa, M., Currier, T., Thiagarajan, M., et al. (2003). TM4: a free, open-source system for microarray data management and analysis. Biotechniques *34*, 374-378.

Sahai, E., and Marshall, C. J. (2002). ROCK and Dia have opposing effects on adherens junctions downstream of Rho. Nat Cell Biol *4*, 408-415.

Saitou, M., Furuse, M., Sasaki, H., Schulzke, J. D., Fromm, M., Takano, H., Noda, T., and Tsukita, S. (2000). Complex phenotype of mice lacking occludin, a component of tight junction strands. Mol Biol Cell *11*, 4131-4142.

Sakakibara, A., Furuse, M., Saitou, M., Ando-Akatsuka, Y., and Tsukita, S. (1997). Possible involvement of phosphorylation of occludin in tight junction formation. J Cell Biol *137*, 1393-1401.

Santin, A. D., Cane, S., Bellone, S., Palmieri, M., Siegel, E. R., Thomas, M., Roman, J. J., Burnett, A., Cannon, M. J., and Pecorelli, S. (2005). Treatment of chemotherapy-resistant human ovarian cancer xenografts in C.B-17/SCID mice by intraperitoneal administration of Clostridium perfringens enterotoxin. Cancer Res *65*, 4334-4342.

Sasaki, H. (2003). Freeze-fracture analysis of renal-epithelial tight junctions. Methods Mol Med *86*, 155-166.

Sauthoff, H., Pipiya, T., Heitner, S., Chen, S., Norman, R. G., Rom, W. N., and Hay, J. G. (2002). Late expression of p53 from a replicating adenovirus improves tumor cell killing and is more tumor cell specific than expression of the adenoviral death protein. Hum Gene Ther *13*, 1859-1871.

# REFERENCES

Scherer, W. F., Syverton, J. T., and Gey, G. O. (1953). Studies on the propagation in vitro of poliomyelitis viruses. IV. Viral multiplication in a stable strain of human malignant epithelial cells (strain HeLa) derived from an epidermoid carcinoma of the cervix. J Exp Med 97, 695-710.

Schneiders, F. I., Maertens, B., Bose, K., Li, Y., Brunken, W. J., Paulsson, M., Smyth, N., and Koch, M. (2007). Binding of netrin-4 to laminin short arms regulates basement membrane assembly. J Biol Chem 282, 23750-23758.

Seidman, J. D., Horkayne-Szakaly, I., Haiba, M., Boice, C. R., Kurman, R. J., and Ronnett, B. M. (2004). The histologic type and stage distribution of ovarian carcinomas of surface epithelial origin. Int J Gynecol Pathol 23, 41-44.

Seth, P., Fitzgerald, D. J., Willingham, M. C., and Pastan, I. (1984a). Role of a low-pH environment in adenovirus enhancement of the toxicity of a Pseudomonas exotoxin-epidermal growth factor conjugate. J Virol 51, 650-655.

Seth, P., Willingham, M. C., and Pastan, I. (1984b). Adenovirus-dependent release of 51Cr from KB cells at an acidic pH. J Biol Chem 259, 14350-14353.

Sevick, E. M., and Jain, R. K. (1991). Effect of red blood cell rigidity on tumor blood flow: increase in viscous resistance during hyperglycemia. Cancer Res 51, 2727-2730.

Shayakhmetov, D. M., Li, Z. Y., Gaggar, A., Gharwan, H., Ternovoi, V., Sandig, V., and Lieber, A. (2004). Genome size and structure determine efficiency of postinternalization steps and gene transfer of capsid-modified adenovirus vectors in a cell-type-specific manner. J Virol 78, 10009-10022.

Shayakhmetov, D. M., Li, Z. Y., Ni, S., and Lieber, A. (2002). Targeting of adenovirus vectors to tumor cells does not enable efficient transduction of breast cancer metastases. Cancer Res 62, 1063-1068.

Shayakhmetov, D. M., Papayannopoulou, T., Stamatoyannopoulos, G., and Lieber, A. (2000). Efficient gene transfer into human CD34(+) cells by a retargeted adenovirus vector. J Virol 74, 2567-2583.

Sheehan, K. M., Gulmann, C., Eichler, G. S., Weinstein, J. N., Barrett, H. L., Kay, E. W., Conroy, R. M., Liotta, L. A., and Petricoin, E. F., 3rd (2008). Signal pathway profiling of epithelial and stromal compartments of colonic carcinoma reveals epithelial-mesenchymal transition. Oncogene 27, 323-331.

Shen, Y. (2007). Viral vectors and their applications. In Fields Virology, 539-564.

Sherwood, O. D. (2004). Relaxin's physiological roles and other diverse actions. Endocr Rev 25, 205-234.

Shinozaki, K., Suominen, E., Carrick, F., Sauter, B., Kahari, V. M., Lieber, A., Woo, S. L., and Savontaus, M. (2006). Efficient infection of tumor endothelial cells by a capsid-modified adenovirus. Gene Ther 13, 52-59.

Silverberg, S. G. (2000). Histopathologic grading of ovarian carcinoma: a review and proposal. Int J Gynecol Pathol 19, 7-15.

Singer, G., Oldt, R., 3rd, Cohen, Y., Wang, B. G., Sidransky, D., Kurman, R. J., and Shih Ie, M. (2003). Mutations in BRAF and KRAS characterize the development of low-grade ovarian serous carcinoma. J Natl Cancer Inst 95, 484-486.

Singh, R., Tian, B., and Kostarelos, K. (2008). Artificial envelopment of nonenveloped viruses: enhancing adenovirus tumor targeting in vivo. FASEB J 22, 3389-3402.

Small, E. J., Carducci, M. A., Burke, J. M., Rodriguez, R., Fong, L., van Ummersen, L., Yu, D. C., Aimi, J., Ando, D., Working, P., et al. (2006). A phase I trial of intravenous CG7870, a replication-selective, prostate-

# References

specific antigen-targeted oncolytic adenovirus, for the treatment of hormone-refractory, metastatic prostate cancer. Mol Ther *14*, 107-117.

Smith, J. G., and Nemerow, G. R. (2008). Mechanism of adenovirus neutralization by Human alpha-defensins. Cell Host Microbe *3*, 11-19.

Smyth, G. K. (2004). Linear models and empirical bayes methods for assessing differential expression in microarray experiments. Stat Appl Genet Mol Biol *3*, Article3.

Soto, E., Yanagisawa, M., Marlow, L. A., Copland, J. A., Perez, E. A., and Anastasiadis, P. Z. (2008). p120 catenin induces opposing effects on tumor cell growth depending on E-cadherin expression. J Cell Biol *183*, 737-749.

Southam, C. M., and Moore, A. E. (1951). West Nile, Ilheus, and Bunyamwera virus infections in man. Am J Trop Med Hyg *31*, 724-741.

Sova, P., Ren, X. W., Ni, S., Bernt, K. M., Mi, J., Kiviat, N., and Lieber, A. (2004). A tumor-targeted and conditionally replicating oncolytic adenovirus vector expressing TRAIL for treatment of liver metastases. Mol Ther *9*, 496-509.

Spentzos, D., Levine, D. A., Ramoni, M. F., Joseph, M., Gu, X., Boyd, J., Libermann, T. A., and Cannistra, S. A. (2004). Gene expression signature with independent prognostic significance in epithelial ovarian cancer. J Clin Oncol *22*, 4700-4710.

Stefani, G., and Slack, F. J. (2008). Small non-coding RNAs in animal development. Nat Rev Mol Cell Biol *9*, 219-230.

Steinwaerder, D. S., Carlson, C. A., Otto, D. L., Li, Z. Y., Ni, S., and Lieber, A. (2001). Tumor-specific gene expression in hepatic metastases by a replication-activated adenovirus vector. Nat Med *7*, 240-243.

Stone, D., David, A., Bolognani, F., Lowenstein, P. R., and Castro, M. G. (2000). Viral vectors for gene delivery and gene therapy within the endocrine system. J Endocrinol *164*, 103-118.

Stone, D., Liu, Y., Shayakhmetov, D., Li, Z. Y., Ni, S., and Lieber, A. (2007). Adenovirus-platelet interaction in blood causes virus sequestration to the reticuloendothelial system of the liver. J Virol *81*, 4866-4871.

Strathdee, G. (2002). Epigenetic versus genetic alterations in the inactivation of E-cadherin. Semin Cancer Biol *12*, 373-379.

Sundfeldt, K., Piontkewitz, Y., Ivarsson, K., Nilsson, O., Hellberg, P., Brannstrom, M., Janson, P. O., Enerback, S., and Hedin, L. (1997). E-cadherin expression in human epithelial ovarian cancer and normal ovary. Int J Cancer *74*, 275-280.

Svensson, U. (1985). Role of vesicles during adenovirus 2 internalization into HeLa cells. J Virol *55*, 442-449.

Taghian, A. G., Abi-Raad, R., Assaad, S. I., Casty, A., Ancukiewicz, M., Yeh, E., Molokhia, P., Attia, K., Sullivan, T., Kuter, I., *et al.* (2005). Paclitaxel decreases the interstitial fluid pressure and improves oxygenation in breast cancers in patients treated with neoadjuvant chemotherapy: clinical implications. J Clin Oncol *23*, 1951-1961.

Tassi, R. A., Bignotti, E., Falchetti, M., Ravanini, M., Calza, S., Ravaggi, A., Bandiera, E., Facchetti, F., Pecorelli, S., and Santin, A. D. (2008). Claudin-7 expression in human epithelial ovarian cancer. Int J Gynecol Cancer *18*, 1262-1271.

Tavazoie, S. F., Alarcon, C., Oskarsson, T., Padua, D., Wang, Q., Bos, P. D., Gerald, W. L., and Massague, J. (2008). Endogenous human microRNAs that suppress breast cancer metastasis. Nature *451*, 147-152.

# References

Thiery, J. P. (2002). Epithelial-mesenchymal transitions in tumour progression. Nat Rev Cancer 2, 442-454.

Thiery, J. P., and Sleeman, J. P. (2006). Complex networks orchestrate epithelial-mesenchymal transitions. Nat Rev Mol Cell Biol 7, 131-142.

Thompson, E. W., Newgreen, D. F., and Tarin, D. (2005). Carcinoma invasion and metastasis: a role for epithelial-mesenchymal transition? Cancer Res 65, 5991-5995; discussion 5995.

Tollefson, A. E., Ryerse, J. S., Scaria, A., Hermiston, T. W., and Wold, W. S. (1996). The E3-11.6-kDa adenovirus death protein (ADP) is required for efficient cell death: characterization of cells infected with adp mutants. Virology 220, 152-162.

Tsukita, S., Yamazaki, Y., Katsuno, T., and Tamura, A. (2008). Tight junction-based epithelial microenvironment and cell proliferation. Oncogene 27, 6930-6938.

Turksen, K., and Troy, T. C. (2004). Barriers built on claudins. J Cell Sci 117, 2435-2447.

Tuve, S., Liu, Y., Tragoolpua, K., Jacobs, J. D., Yumul, R. C., Li, Z. Y., Strauss, R., Hellstrom, K. E., Disis, M. L., Roffler, S., and Lieber, A. (2009). In situ adenovirus vaccination engages T effector cells against cancer. Vaccine 27, 4225-4239.

Tuve, S., Wang, H., Jacobs, J. D., Yumul, R. C., Smith, D. F., and Lieber, A. (2008). Role of cellular heparan sulfate proteoglycans in infection of human adenovirus serotype 3 and 35. PLoS Pathog 4, e1000189.

Tuve, S., Wang, H., Ware, C., Liu, Y., Gaggar, A., Bernt, K., Shayakhmetov, D., Li, Z., Strauss, R., Stone, D., and Lieber, A. (2006). A new group B adenovirus receptor is expressed at high levels on human stem and tumor cells. J Virol 80, 12109-12120.

Uchida, N., Honjo, Y., Johnson, K. R., Wheelock, M. J., and Takeichi, M. (1996). The catenin/cadherin adhesion system is localized in synaptic junctions bordering transmitter release zones. J Cell Biol 135, 767-779.

Vaezi, A., Bauer, C., Vasioukhin, V., and Fuchs, E. (2002). Actin cable dynamics and Rho/Rock orchestrate a polarized cytoskeletal architecture in the early steps of assembling a stratified epithelium. Dev Cell 3, 367-381.

Van Itallie, C. M., Fanning, A. S., and Anderson, J. M. (2003). Reversal of charge selectivity in cation or anion-selective epithelial lines by expression of different claudins. Am J Physiol Renal Physiol 285, F1078-1084.

van Raaij, M. J., Chouin, E., van der Zandt, H., Bergelson, J. M., and Cusack, S. (2000). Dimeric structure of the coxsackievirus and adenovirus receptor D1 domain at 1.7 A resolution. Structure 8, 1147-1155.

Vandewalle, C., Comijn, J., De Craene, B., Vermassen, P., Bruyneel, E., Andersen, H., Tulchinsky, E., Van Roy, F., and Berx, G. (2005). SIP1/ZEB2 induces EMT by repressing genes of different epithelial cell-cell junctions. Nucleic Acids Res 33, 6566-6578.

Vasey, P. A., Shulman, L. N., Campos, S., Davis, J., Gore, M., Johnston, S., Kirn, D. H., O'Neill, V., Siddiqui, N., Seiden, M. V., and Kaye, S. B. (2002). Phase I trial of intraperitoneal injection of the E1B-55-kd-gene-deleted adenovirus ONYX-015 (dl1520) given on days 1 through 5 every 3 weeks in patients with recurrent/refractory epithelial ovarian cancer. J Clin Oncol 20, 1562-1569.

Vega, F. M., and Ridley, A. J. (2008). Rho GTPases in cancer cell biology. FEBS Lett 582, 2093-2101.

Venkiteswaran, K., Xiao, K., Summers, S., Calkins, C. C., Vincent, P. A., Pumiglia, K., and Kowalczyk, A. P. (2002). Regulation of endothelial barrier function and growth by VE-cadherin, plakoglobin, and beta-catenin. Am J Physiol Cell Physiol 283, C811-821.

# References

Vigant, F., Descamps, D., Jullienne, B., Esselin, S., Connault, E., Opolon, P., Tordjmann, T., Vigne, E., Perricaudet, M., and Benihoud, K. (2008). Substitution of hexon hypervariable region 5 of adenovirus serotype 5 abrogates blood factor binding and limits gene transfer to liver. Mol Ther *16*, 1474-1480.

Vincan, E., and Barker, N. (2008). The upstream components of the Wnt signalling pathway in the dynamic EMT and MET associated with colorectal cancer progression. Clin Exp Metastasis *25*, 657-663.

Vogelstein, B., and Kinzler, K. W. (2004). Cancer genes and the pathways they control. Nat Med *10*, 789-799.

Von Seggern, D. J., Chiu, C. Y., Fleck, S. K., Stewart, P. L., and Nemerow, G. R. (1999). A helper-independent adenovirus vector with E1, E3, and fiber deleted: structure and infectivity of fiberless particles. J Virol *73*, 1601-1608.

Voutilainen, K. A., Anttila, M. A., Sillanpaa, S. M., Ropponen, K. M., Saarikoski, S. V., Juhola, M. T., and Kosma, V. M. (2006). Prognostic significance of E-cadherin-catenin complex in epithelial ovarian cancer. J Clin Pathol *59*, 460-467.

Waddington, S. N., McVey, J. H., Bhella, D., Parker, A. L., Barker, K., Atoda, H., Pink, R., Buckley, S. M., Greig, J. A., Denby, L., et al. (2008). Adenovirus serotype 5 hexon mediates liver gene transfer. Cell *132*, 397-409.

Wagner, O. I., Rammensee, S., Korde, N., Wen, Q., Leterrier, J. F., and Janmey, P. A. (2007). Softness, strength and self-repair in intermediate filament networks. Exp Cell Res *313*, 2228-2235.

Walters, R. W., Freimuth, P., Moninger, T. O., Ganske, I., Zabner, J., and Welsh, M. J. (2002). Adenovirus fiber disrupts CAR-mediated intercellular adhesion allowing virus escape. Cell *110*, 789-799.

Walters, R. W., Grunst, T., Bergelson, J. M., Finberg, R. W., Welsh, M. J., and Zabner, J. (1999). Basolateral localization of fiber receptors limits adenovirus infection from the apical surface of airway epithelia. J Biol Chem *274*, 10219-10226.

Wenning, L. A., and Murphy, R. M. (1999). Coupled cellular trafficking and diffusional limitations in delivery of immunotoxins to multicell tumor spheroids. Biotechnol Bioeng *62*, 562-575.

Wickham, T. J., Filardo, E. J., Cheresh, D. A., and Nemerow, G. R. (1994). Integrin alpha v beta 5 selectively promotes adenovirus mediated cell membrane permeabilization. J Cell Biol *127*, 257-264.

Wickham, T. J., Mathias, P., Cheresh, D. A., and Nemerow, G. R. (1993). Integrins alpha v beta 3 and alpha v beta 5 promote adenovirus internalization but not virus attachment. Cell *73*, 309-319.

Wiethoff, C. M., Wodrich, H., Gerace, L., and Nemerow, G. R. (2005). Adenovirus protein VI mediates membrane disruption following capsid disassembly. J Virol *79*, 1992-2000.

Wiley, S. R., Schooley, K., Smolak, P. J., Din, W. S., Huang, C. P., Nicholl, J. K., Sutherland, G. R., Smith, T. D., Rauch, C., Smith, C. A., and et al. (1995). Identification and characterization of a new member of the TNF family that induces apoptosis. Immunity *3*, 673-682.

Wisse, E., Jacobs, F., Topal, B., Frederik, P., and De Geest, B. (2008). The size of endothelial fenestrae in human liver sinusoids: implications for hepatocyte-directed gene transfer. Gene Ther *15*, 1193-1199.

Wohlfahrt, M. E., Beard, B. C., Lieber, A., and Kiem, H. P. (2007). A capsid-modified, conditionally replicating oncolytic adenovirus vector expressing TRAIL Leads to enhanced cancer cell killing in human glioblastoma models. Cancer Res *67*, 8783-8790.

Wojciak-Stothard, B., Potempa, S., Eichholtz, T., and Ridley, A. J. (2001). Rho and Rac but not Cdc42 regulate endothelial cell permeability. J Cell Sci *114*, 1343-1355.

# REFERENCES

Wold, W. S., Doronin, K., Toth, K., Kuppuswamy, M., Lichtenstein, D. L., and Tollefson, A. E. (1999). Immune responses to adenoviruses: viral evasion mechanisms and their implications for the clinic. Curr Opin Immunol *11*, 380-386.

Wong, A. S., and Auersperg, N. (2002). Normal ovarian surface epithelium. Cancer Treat Res *107*, 161-183.

Wong, A. S., and Gumbiner, B. M. (2003). Adhesion-independent mechanism for suppression of tumor cell invasion by E-cadherin. J Cell Biol *161*, 1191-1203.

Xiao, K., Allison, D. F., Buckley, K. M., Kottke, M. D., Vincent, P. A., Faundez, V., and Kowalczyk, A. P. (2003). Cellular levels of p120 catenin function as a set point for cadherin expression levels in microvascular endothelial cells. J Cell Biol *163*, 535-545.

Xie, X., Zhao, X., Liu, Y., Young, C. Y., Tindall, D. J., Slawin, K. M., and Spencer, D. M. (2001). Robust prostate-specific expression for targeted gene therapy based on the human kallikrein 2 promoter. Hum Gene Ther *12*, 549-561.

Yamada, S., Pokutta, S., Drees, F., Weis, W. I., and Nelson, W. J. (2005). Deconstructing the cadherin-catenin-actin complex. Cell *123*, 889-901.

Yamamoto, M., Alemany, R., Adachi, Y., Grizzle, W. E., and Curiel, D. T. (2001). Characterization of the cyclooxygenase-2 promoter in an adenoviral vector and its application for the mitigation of toxicity in suicide gene therapy of gastrointestinal cancers. Mol Ther *3*, 385-394.

Yanagisawa, M., Huveldt, D., Kreinest, P., Lohse, C. M., Cheville, J. C., Parker, A. S., Copland, J. A., and Anastasiadis, P. Z. (2008). A p120 catenin isoform switch affects Rho activity, induces tumor cell invasion, and predicts metastatic disease. J Biol Chem *283*, 18344-18354.

Yang, J., Mani, S. A., Donaher, J. L., Ramaswamy, S., Itzykson, R. A., Come, C., Savagner, P., Gitelman, I., Richardson, A., and Weinberg, R. A. (2004). Twist, a master regulator of morphogenesis, plays an essential role in tumor metastasis. Cell *117*, 927-939.

Yang, J., and Weinberg, R. A. (2008). Epithelial-mesenchymal transition: at the crossroads of development and tumor metastasis. Dev Cell *14*, 818-829.

Yap, A. S., Niessen, C. M., and Gumbiner, B. M. (1998). The juxtamembrane region of the cadherin cytoplasmic tail supports lateral clustering, adhesive strengthening, and interaction with p120ctn. J Cell Biol *141*, 779-789.

Yates, C. C., Shepard, C. R., Stolz, D. B., and Wells, A. (2007). Co-culturing human prostate carcinoma cells with hepatocytes leads to increased expression of E-cadherin. Br J Cancer *96*, 1246-1252.

Yew, P. R., and Berk, A. J. (1992). Inhibition of p53 transactivation required for transformation by adenovirus early 1B protein. Nature *357*, 82-85.

Yoshida, C., and Takeichi, M. (1982). Teratocarcinoma cell adhesion: identification of a cell-surface protein involved in calcium-dependent cell aggregation. Cell *28*, 217-224.

Yu, W., and Fang, H. (2007). Clinical trials with oncolytic adenovirus in China. Curr Cancer Drug Targets *7*, 141-148.

Zallen, J. A. (2007). Planar polarity and tissue morphogenesis. Cell *129*, 1051-1063.

Zhi, Y., Figueredo, J., Kobinger, G. P., Hagan, H., Calcedo, R., Miller, J. R., Gao, G., and Wilson, J. M. (2006). Efficacy of severe acute respiratory syndrome vaccine based on a nonhuman primate adenovirus in the presence of immunity against human adenovirus. Hum Gene Ther *17*, 500-506.

## References

Zochowska, M., Paca, A., Schoehn, G., Andrieu, J. P., Chroboczek, J., Dublet, B., and Szolajska, E. (2009). Adenovirus dodecahedron, as a drug delivery vector. PLoS One 4, e5569.

# 7. Supplement

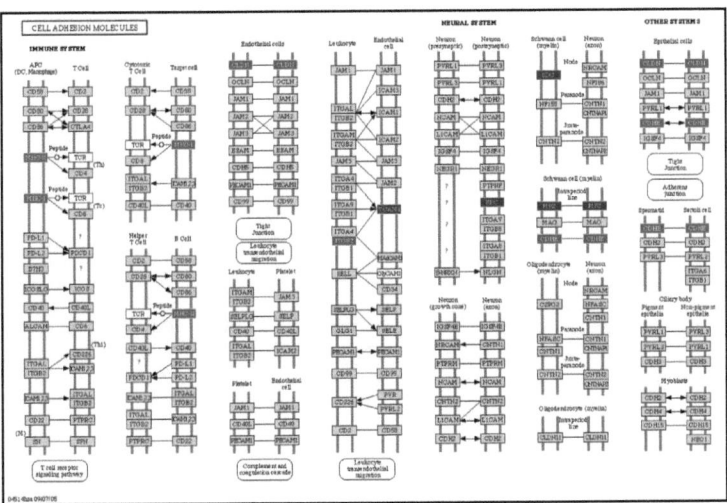

**Supplementary Figure 1:** Cell adherens proteins that were differentially expressed in S/M and R/E cells. Shown is the status in R/E cells. Significantly up-regulated genes are red, genes that were down-regulated appear in blue.

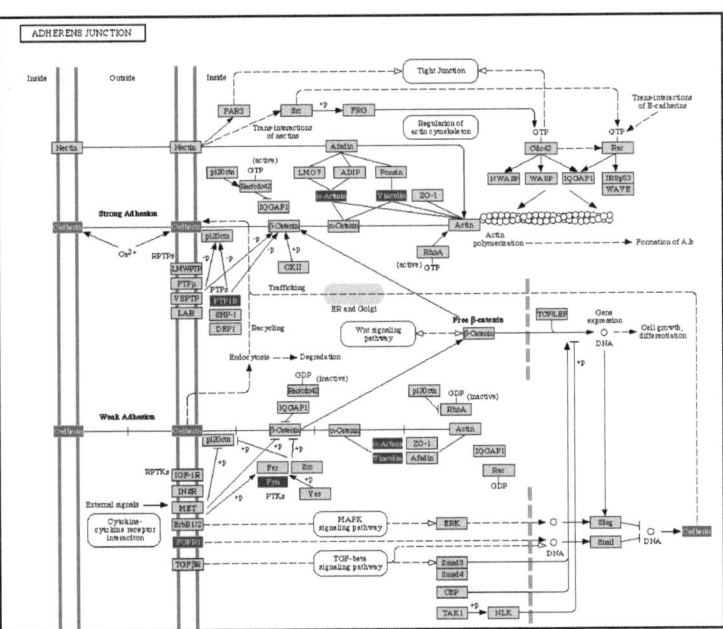

**Supplementary Figure 2:** Adherens junction proteins that were differentially expressed in S/M and R/E cells. Shown is the status in R/E cells. Significantly up-regulated genes are red, genes that were down-regulated appear in blue.

## SUPPLEMENT

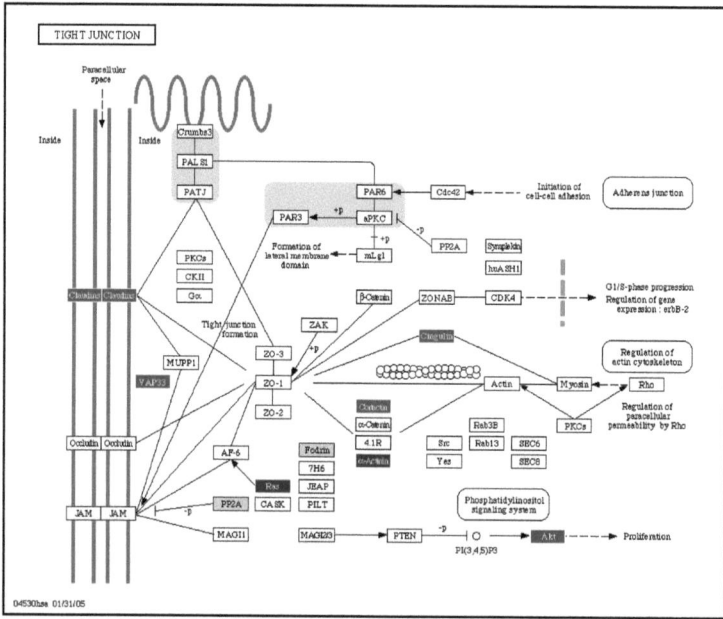

**Supplementary Figure 3: Tight junction proteins that were differentially expressed in S/M and R/E cells.** Shown is the status in R/E cells. Significantly up-regulated genes are red, genes that were down-regulated appear in blue.

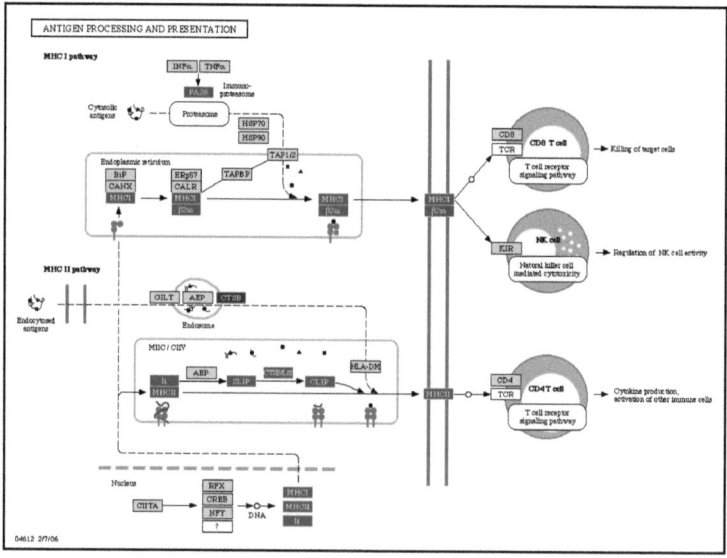

**Supplementary Figure 4: Differentially expressed proteins involved in antigen presentation.** Compared are R/E and S/M cells, shown for R/E. Significantly up-regulated genes are red, genes that were down-regulated appear in blue.

## SUPPLEMENT

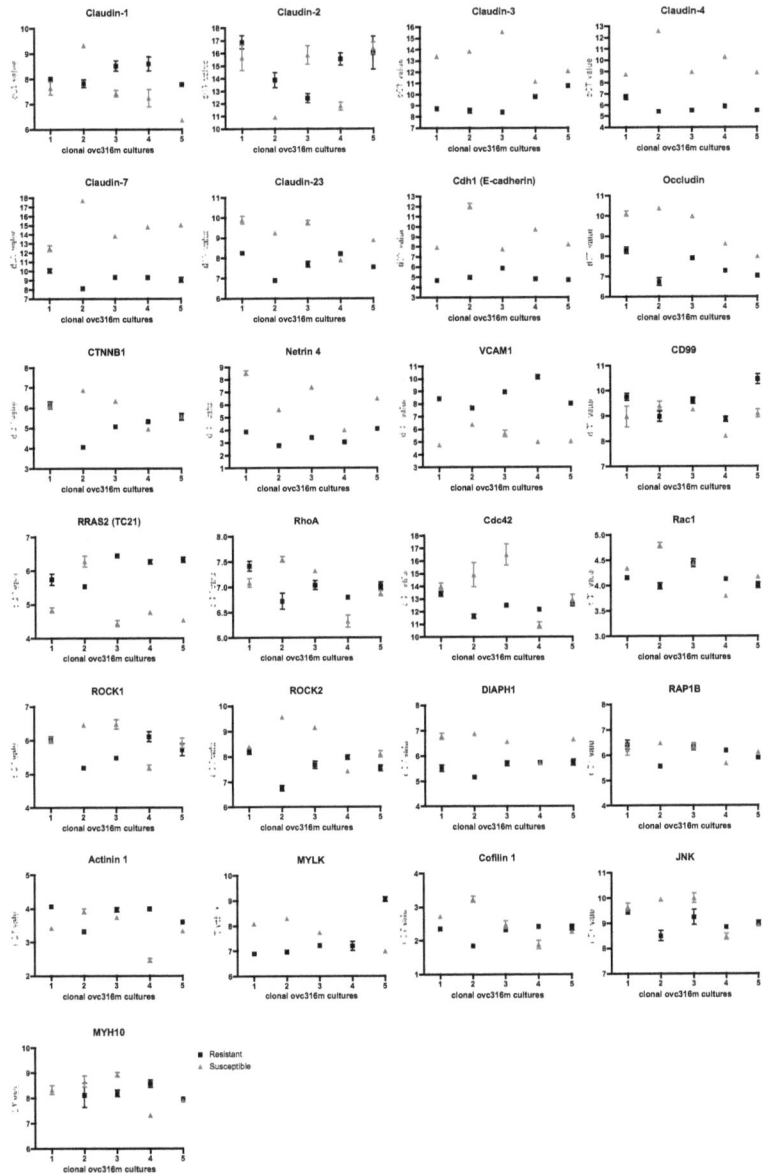

**Supplementary Figure 5: Expression status of proteins involved in cell adherens pathways.** qRT-PCR data of R/E and S/M cultures (N=5 each) visualized in deltaCT values. Technical standard deviations are included, but mostly so small that not visible. Data are normalized to expression of housekeeping gene GAPDH. Results after normalization to Cyclophilin A did not significantly differ. Resistant clones are shown in black, susceptible cultures are grey.

## Acknowledgements

First of all, I want to thank Professor André Lieber for the opportunity to work in his laboratory at the University of Washington. The past five years under his supervision have been extremely beneficial for my general scientific understanding. I deeply thank him especially for his enormous trust in my abilities, his support, and his motivation, especially in tough phases of the project. He gave me many opportunities to present my data at scientific conferences, always had an open door, and granted a lot of freedom to develop my own ideas.

Furthermore, I want to thank Professor Wolfgang Uckert for the supervision of my doctoral thesis and for many interesting scientific or non-scientific conversations during my regular visits in Berlin, which I always enjoyed. I am thankful to Professor Wolfgang Lockau, Professor Detlef Krüger, Professor Jiri Bartek, and Dr. Günter Cichon for the critical review of my thesis and for being part of my committee.

I am grateful to all members of the Lieber lab for providing a very inspiring and pleasant working environment: Hongjie Wang, Ying Liu, Nelson Di Paolo, Dmitry Shayakhmetov, Anuj Gaggar, Roma Yumul, Jeff Jacobs, and Shaoheng Ni. Specifically, I thank our "lab mum" Zong-Yi Li for introducing me into the world of immunohistochemistry and Sebastian Tuve for countless enjoyable latte breaks and a lot of good times in the lab. Most of all I want to thank Daniel Stone for his guidance, his open ears to all sorts of problems and for his friendship (I will bring my suit to the Caribbean for you!). Moreover, I want to acknowledge Pavel Sova, who was a great help for the DNA expression array analysis and therefore had immense impact on the study. Special thanks go to Sari Pesonen and Akseli Hemminki from the University of Helskinki, who gave me the opportunity for a fruitful scientific collaboration.

I also say "thank you" to my scientific friends Paul Brinkkoetter, Thomas Möller, Markus Welcker, and Martin Wohlfahrt. All of them showed great interest in my study and managed to get me out of the lab, whenever it was necessary. Furthermore, I want to acknowledge my soccer team "Bottoms Up!". You guys have made all of my Monday evenings a special highlight of the week. A great thank-you goes to my dear roomies Jackie and Frank and to my good friend Aune (who even managed to turn me into a runner). All of you gave me a warm family feeling and turned my stay in Seattle into a great time. I am also extremely thankful for the intensive contact with my friends in Europe: Philipp, Robert, Frank, Stefan, Pierre, Martin, Jule, Ulla, Feliks, Melanie, Moritz and Carsten. Finally, I want to thank my mum Irene Strauss, my grandma Brigitte Strauss and the rest of my family for their enormous support.

I dedicate this thesis to my deeply missed father Michael Strauss, who sparked my interest in science when I was a little boy.

Robert Strauss,

July 2009

## I want morebooks!

Buy your books fast and straightforward online - at one of world's fastest growing online book stores! Environmentally sound due to Print-on-Demand technologies.

Buy your books online at
**www.morebooks.shop**

Kaufen Sie Ihre Bücher schnell und unkompliziert online – auf einer der am schnellsten wachsenden Buchhandelsplattformen weltweit! Dank Print-On-Demand umwelt- und ressourcenschonend produziert.

Bücher schneller online kaufen
**www.morebooks.shop**

KS OmniScriptum Publishing
Brivibas gatve 197
LV-1039 Riga, Latvia
Telefax +371 686 204 55

info@omniscriptum.com
www.omniscriptum.com

Printed by Books on Demand GmbH, Norderstedt / Germany